REA's Books Are The Best...
They have rescued lots of grades and more!

(a sample of the <u>hundreds of letters</u> REA receives each year)

" Your books are great! They are very helpful, and have upped my grade in every class. Thank you for such a great product. "

Student, Seattle, WA

" Your book has really helped me sharpen my skills and improve my weak areas. Definitely will buy more. "

Student, Buffalo, NY

" Compared to the other books that my fellow students had, your book was the most useful in helping me get a great score. "

Student, North Hollywood, CA

" I really appreciate the help from your excellent book. Please keep up your great work. "

Student, Albuquerque, NM

" Your book was such a better value and was so much more complete than anything your competition has produced (and I have them all)! "

Teacher, Virginia Beach, VA

(more on next page)

(continued from previous page)

"Your books have saved my GPA, and quite possibly my sanity. My course grade is now an 'A', and I couldn't be happier."
Student, Winchester, IN

"These books are the best review books on the market. They are fantastic!"
Student, New Orleans, LA

"Your book was responsible for my success on the exam. . . I will look for REA the next time I need help."
Student, Chesterfield, MO

"I think it is the greatest study guide I have ever used!"
Student, Anchorage, AK

"I encourage others to buy REA because of their superiority. Please continue to produce the best quality books on the market."
Student, San Jose, CA

"Just a short note to say thanks for the great support your book gave me in helping me pass the test . . . I'm on my way to a B.S. degree because of you !"
Student, Orlando, FL

PHYSICS

By the Staff of
Research & Education Association
Dr. M. Fogiel, Director

Research & Education Association
61 Ethel Road West
Piscataway, New Jersey 08854

SUPER REVIEW®
OF PHYSICS

Year 2003 Printing

Printed in the United States of America

Library of Congress Control Number 00-131304

International Standard Book Number 0-87891-087-5

SUPER REVIEW is a registered trademark of
Research & Education Association, Piscataway, New Jersey 08854

I-4

WHAT THIS Super Review WILL DO FOR YOU

This **Super Review** provides all that you need to know to do your homework effectively and succeed on exams and quizzes.

The book focuses on the core aspects of the subject, and helps you to grasp the important elements quickly and easily.

Outstanding **Super Review** features:

- Topics are covered in logical sequence

- Topics are reviewed in a concise and comprehensive manner

- The material is presented in student-friendly language that makes it easy to follow and understand

- Individual topics can be easily located

- Provides excellent preparation for midterms, finals and in-between quizzes

- In every chapter, reviews of individual topics are accompanied by Questions **Q** and Answers **A** that show how to work out specific problems

- At the end of most chapters, quizzes with answers are included to enable you to practice and test yourself to pinpoint your strengths and weaknesses

- Written by professionals and test experts who function as your very own tutors

Dr. Max Fogiel
Program Director

CONTENTS

19 ENTROPY AND THE SECOND LAW OF THERMODYNAMICS

20 ELECTROMAGNETISM AND ELECTRIC FIELDS

21 GAUSS' LAW

22 ELECTRIC POTENTIAL (VOLTAGE)

23 CAPACITORS

30 ALTERNATING CURRENTS

31 ELECTROMAGNETIC WAVES

32 GEOMETRICAL OPTICS

33   INTERFERENCE

34   DIFFRACTION

CHAPTER 1

Vectors and Scalars

1.1 Basic Definitions of Vectors and Scalars

a) A vector is a quantity that has both magnitude and direction.

Some typical vector quantities are displacement, velocity, force, acceleration, momentum, electric field strength, and magnetic field strength.

b) A scalar is a quantity that has magnitude but no direction.

Some typical scalar quantities are mass, length, time, density, energy, and temperature.

Problem Solving Example:

Find the resultant of the vectors s_1 and s_2 specified in the figures below.

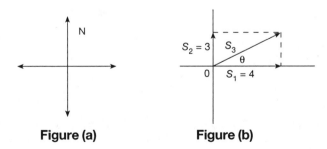

Figure (a)   Figure (b)

A From the Pythagorean theorem, $s_1^2 + s_2^2 = s_3^2$, or $4^2 + 3^2 = s_3^2$, and so we get $s_3 = 5$ units. The direction of S_3 may be specified by the angle θ which it makes with s_1.

$$\sin \theta = \frac{s_2}{s_3} = 0.60 \text{ gives } \theta = 37°.$$

Resultant s_3 therefore represents a displacement of 5 units from 0 in the direction 37° north of east.

1.2 Addition of Vectors (a– + b–) —Geometric Methods

a) Triangle Method (Head-to-Tail Method)

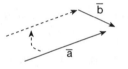 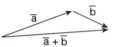

(i) Attach the head of $\bar{a}$ to the tail of $\bar{b}$

(ii) By connecting the head of $\bar{a}$ to the tail of $\bar{b}$, the vector $\bar{a} + \bar{b}$ is defined.

Figure 1.1 Triangle Method of Adding Vectors

b) Parallelogram Method (Tail-to-Tail Method)

(i) Join the tails of the two vectors.

(ii) Construct a parallelogram having $\bar{a}$ and $\bar{b}$ as two of its sides. The diagonal of the parallelogram repre-sents the vector $\bar{a} + \bar{b}$.

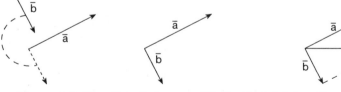

Figure 1.2 The Parallelogram Method of Adding Vectors

Problem Solving Example:

A ship leaving its port sails due north for 30 kms and then 50 kms in a direction 60° east of north. See the figure below. At the end of this time where is the ship relative to its port?

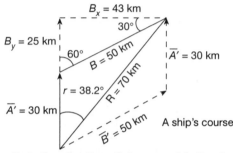

A ship's course

Solution by Parallelogram Method:

The figure shows the parallelogram completed by the dashed vectors $\overline{A}'$ and $\overline{B}'$. Also shown is the resultant $\overline{R}$, which is found to represent about 70 km . Angle r is found to be about 38.2° east of north.

Solution by Component Method:

The figure also shows the vector $\overline{B}$ resolved into the components $\overline{B}_X$ and $\overline{B}_Y$, which are found to be 43 km and 25 km, respectively. By trigonometry

$$\overrightarrow{B}_X = 50 \text{ km} \times \cos 30° = 43 \text{ km},$$

and

$$\overrightarrow{B}_Y = 50 \text{ km} \times \sin 30° = 25 \text{ km},$$

Since $\overline{A}'$ and $\overline{B}'$ lie along the same direction in this problem, we add them directly to get 30 km + 25 km, or 55 km. We then have a right triangle with one side equal to 55 km and the other side equal to 43 km. From these data we find the resultant R according to the equation:

$$R^2 = 55^2 + 43^2$$

$$R = \text{about 70 km}$$

Solution by the Cosine Law:

In solving this problem by means of the cosine law, we write

$$R^2 = A^2 + B^2 + 2AB \cos \theta$$

$$F^2 = 30^2 + 50^2 + 2 \times 30 \times 50 \times 0.5000$$

$$= 4,900$$

hence the magnitude of R is

$$R = 70 \text{ km}$$

$$\tan R = \frac{B \sin \theta}{A + B \cos \theta} = \frac{50 \times 0.866}{30 + 50 \times 0.500} = 0.788$$

hence
$$R = 38.2° \text{ approximately.}$$

1.3   Subtraction of Vectors

The subtraction of a vector is defined as the addition of the corresponding negative vector. Therefore, the vector $\overline{P} - \overline{F}$ is obtained by adding the vector $(-\overline{F})$ to the vector $\overline{P}$, i.e., $\overline{P} + (-\overline{F})$. (See Figure 1.3.)

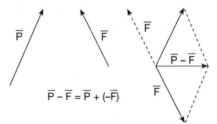

Figure 1.3 The Subtraction of a Vector

Problem Solving Example:

 The crew of a spacecraft, which is out in space with the rocket motors switched off, experience no weight and can therefore glide through the air inside the craft.

The cabin of such a spaceship is cube of side 5 m. An astronaut working in one corner requires a tool which is in a cupboard in the diametrically opposite corner of the cabin. What is the minimum distance that he has to glide and at what angle to the floor must he launch himself?

If he decided instead to put on boots with magnetic soles, which allow him to remain fixed to the metal of the cabin, and thus enable him to walk along the floor and, in the absence of gravitational effects, up the walls and across the ceiling, what is the minimum distance he needs to get to the cupboard?

 Figure (a) shows the cabin. Axes have been set up with the x, y, and z directions coinciding with the length, breadth, and height of the room. The astronaut must get from point A to point B. The vector $\overline{A}$ going from the origin 0 to point A is

$$\overline{A} = (5 \text{ m}, 0, 0)$$

The vector from 0 to point B is:

$$B = (0, 5 \text{ m}, 5 \text{ m})$$

The vector going from A to B is then:

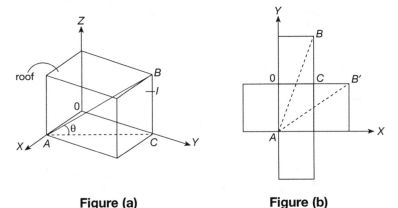

Figure (a)   **Figure (b)**

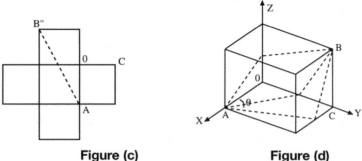

Figure (c)   **Figure (d)**

$$B - A = (0, 5 \text{ m}, 5 \text{ m}) - (5 \text{ m}, 0, 0)$$

$$= (-5 \text{ m}, 5 \text{ m}, 5 \text{ m})$$

Its length is:

$$\left| \overline{B} - \overline{A} \right| = \sqrt{(-5 \text{ m})^2 + (5 \text{ m})^2 + (5 \text{ m})^2} = 5\sqrt{3} \text{ m}$$

This is the distance the astronaut must glide.

The angle to the floor at which he launches himself is q, where tan $q = BC/AC$. Point C has coordinates $(0, 5 \text{ m}, 0)$. Thus, BC has length 5 m and AC has length l, breadth b, and height h. They are all different; the three routes correspond to vectors having components $(l; b + h)$, $(b; l + h)$, and $(h; l + b)$. The shortest of these will be the one in which the x-component is the longest dimension and the y-component the sum of the other two.

1.4 The Components of a Vector

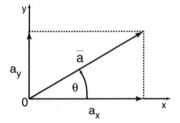

a_x and a_y are the components of a vector $\bar{a}$. The angle θ is measured counterclockwise from the positive x-axis. The components are formed when we draw perpendicular lines to the chosen axes.

Figure 1.4 The Formation of Vector Components on the Positive y Axis

a) The components of a vector are given by

$$A_x = A\cos\theta$$
$$A_y = A\sin\theta$$

A component is equal to the product of the magnitude of vector A and the cosine of the angle between the positive axis and the vector.

b) The magnitude can be expressed in terms of the components.

$$A = \sqrt{A_x^2 + A_y^2}$$

For the angle θ,

$$\tan\theta = \frac{A_y}{A_x}$$

c) A vector $\overline{F}$ can be written in terms of its components F_x and F_y

$$\overline{F} = \overline{i}F_x + \overline{j}F_y$$

where $\overline{i}$ and $\overline{j}$ represent perpendicular unit vectors (magnitude = 1) along the x- and y-axis.

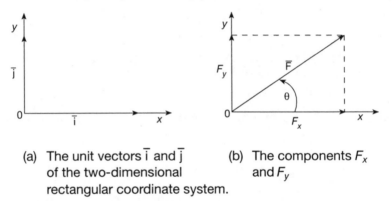

(a) The unit vectors $\overline{i}$ and $\overline{j}$ of the two-dimensional rectangular coordinate system.

(b) The components F_x and F_y

Figure 1.5 Vector Components and the Unit Vector

Problem Solving Examples:

Q Three forces acting at a point are

$$\overline{F}_1 = 2\overline{i} - \overline{j} + 3\overline{k},$$

$$\overline{F}_2 = -\overline{i} + 3\overline{j} + 2\overline{k}$$

and
$$\overline{F}_3 = -\overline{i} + 2\overline{j} - \overline{k}$$

Find the directions and magnitude of

$$\overline{F}_1 + \overline{F}_2 + \overline{F}_3, \quad \overline{F}_1 - \overline{F}_2 + \overline{F}_3, \quad \text{and} \quad \overline{F}_1 + \overline{F}_2 - \overline{F}_3.$$

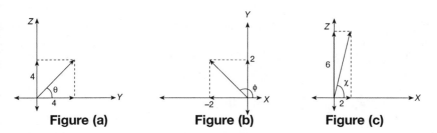

Figure (a)　　　　**Figure (b)**　　　　**Figure (c)**

A When vectors are added (or subtracted), their components in the directions of the unit vectors add (or subtract) algebraically. Thus, since

$$\overline{F}_1 = 2\overline{i} - \overline{j} + 3\overline{k}$$

$$\overline{F}_2 = -\overline{i} + 3\overline{j} + 2\overline{k}$$

$$\overline{F}_3 = -\overline{i} + 2\overline{j} - \overline{k}$$

then it follows that

$$\overline{F}_1 + \overline{F}_2 + \overline{F}_3 = (2-1-1)\overline{i} + (-1+3+2)\overline{j} + (3+2-1)\overline{k}$$

$$= 0\overline{i} + 4\overline{j} + 4\overline{k}$$

Similarly,

$$\overline{F}_1 - \overline{F}_2 + \overline{F}_3 = [2-(-1)-1]\overline{i} + (-1-3+2)\overline{j} + (3-2-1)\overline{k}$$

$$= 2\overline{i} - 2\overline{j} + 0\overline{k}$$

$$\overline{F}_1 + \overline{F}_2 - \overline{F}_3 = [2-(-1)-1]\overline{i} + (-1+3-2)\overline{j} + [3+2-(-1)]\overline{k}$$

$$= 2\overline{i} + 0\overline{j} + 6\overline{k}$$

The vector $\overline{F}_1 + \overline{F}_2 + \overline{F}_3$ thus has no component in the x-direction, one of 4 units in the y-direction, and one of 4 units in the z-direction. It therefore has a magnitude of

$$\sqrt{4^2 + 4^2} \text{ units } + 4\sqrt{2} \text{ units } = 5.66 \text{ units,}$$

and lies in the $y - z$ plane, making an angle θ with the y-axis, as shown in Figure (a), where $\tan \theta = {}^4/_4 = 1$. Thus, $\theta = 45°$.

Similarly, $\overline{F}_1 - \overline{F}_2 + \overline{F}_3$ has a magnitude of

$$2\sqrt{2} \text{ units } = 2.82 \text{ units,}$$

and lies in the $x - y$ plane, making an angle ϕ with the x-axis, as shown in Figure (b), where

$$\tan \phi = \frac{+2}{-2} = -1$$

Thus, $\phi = 315°$.

Also, $\overline{F}_1 + \overline{F}_2 - \overline{F}_3$ has a magnitude of

$$\sqrt{2^2 + 6^2} \text{ units } = 2\sqrt{10} \text{ units } = 6.32,$$

and lies in the $x - z$ plane at an angle χ to the x-axis, as shown in Figure (c), where

$$\tan \chi = \frac{6}{2} = 3$$

Thus, $\chi = 71°34'$.

Q One of the holes on a golf course runs due west. When playing on it recently, a golfer sliced his tee shot badly and landed in a thick rough 120 m WNW of the tee. The ball was in such a bad lie that he was forced to blast it SSW onto the fairway, where it came to rest 75 m from him. A chip shot onto the green, which carried 64 m, took the ball to a point 2 m past the hole on a direct line from hole to tee. He sank the putt. What is the length of this hole? (Assume the golf course to be flat.)

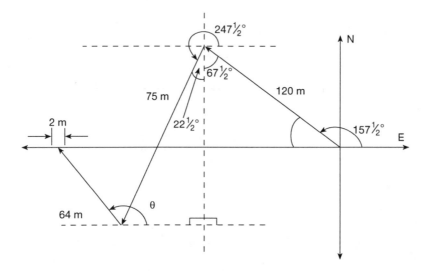

Since the course is flat, all displacements are in the one horizontal plane. Since we know that the hole is due west of the tee, we only need to calculate its easterly component, which is the sum of the easterly components of the ball's displacements (see figure). We take east to be the positive abscissa of the axes shown, and the direction angles clockwise from the positive east axis.

Since∫e know that WNW means $22\frac{1}{2}°$ west of northwest, or $\phi = 157\frac{1}{2}°$, and that SSW means $22\frac{1}{2}°$ south of southwest, or $\phi = 247\frac{1}{2}°$,

Σ easterly components = 120 cos $157\frac{1}{2}°$ m + 75 cos $247\frac{1}{2}°$ m

$$+ 64 \cos \theta \text{ m} + 2 \text{ m}$$

$$= -110.9 \text{ m} - 28.7 \text{ m} + 64 \cos \theta \text{ m} + 2 \text{ m}$$

$$= -137.6 \text{ m} + 64 \cos \theta \text{ m}$$

We can solve for θ by noting that the sum of the northerly components of displacement must equal zero:

Σ northerly components $= 120 \sin 157\frac{1}{2}^\circ \text{ m} + 75 \sin 247\frac{1}{2}^\circ \text{ m}$

$$+ 64 \sin \theta \text{ m} = 0$$

$$\sin \theta = \frac{-120 \sin 157\frac{1}{2}^\circ - 75 \sin 247\frac{1}{2}^\circ}{64}$$

$$= \frac{-45.9 + 69.3}{64} = \frac{23.4}{64}$$

Thus: $\qquad\qquad \theta = 158.6^\circ$

$$\cos \theta = -0.93125$$

Finally, inserting this into the equation for the sum of the easterly components:

Σ easterly components $= -137.6 \text{ m} + 64 (-0.93125) \text{ m}$

$$= -137.6 \text{ m} - 59.6 \text{ m}$$

$$= -197.2 \text{ m}$$

Thus, the hole is 197.2 m due west of the tee.

1.5   The Unit Vector

A unit vector in the direction of a vector $\bar{a}$ is given by

$$\hat{u}_a = \frac{\bar{a}}{|\bar{a}|} = \frac{\bar{a}}{a} = \left[\frac{a_x}{a} \hat{i} + \frac{a_y}{a} \hat{j} \right]$$

Problem Solving Example:

We consider the vector

$$\overline{A} = 3\overline{x} + \overline{y} + 2\overline{z}$$

a) Find the length of $\overline{A}$.

b) What is the length of the projection of $\overline{A}$ on the xy plane?

c) Construct a vector in the xy plane and perpendicular to $\overline{A}$.

d) Construct the unit vector $\overline{B}$.

e) Find the scalar product with $\overline{A}$ of the vector $\overline{C} = 2\overline{X}$.

f) Find the form $\overline{A}$ of and $\overline{C}$ in a reference frame obtained from the old reference frame by a rotation of $\pi/_2$ clockwise looking along the positive z axis.

g) Find the scalar product $\overline{A} \times \overline{C}$ in the primed coordinate system.

h) Find the vector product $\overline{A} \times \overline{C}$.

i) Form the vector $\overline{A} - \overline{C}$.

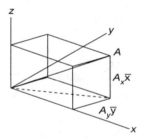

Figure 1

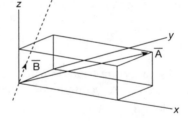

Figure 2

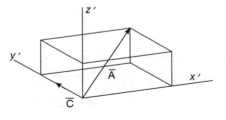

Figure 3

Figure 4

The primed reference frame x', y', z' is generated from the unprimed system x, y, z by a rotation of $+ \pi/2$ about the z-axis.

a) When a vector is given in the form

$$A_x \overline{x} + A_y \overline{y} + A_z \overline{z}$$

its length is given by

$$\sqrt{A_x^2 + A_y^2 + A_2^2}$$

This can be seen from Figure 1 (on previous page). Vector $\overline{A}$ has components in the x, y, and z directions. The x and y components form the legs of a right triangle. By the Pythagorean theorem, the length of the hypotenuse of this triangle is

$$\sqrt{A_x^2 + A_y^2}$$

But this line segment whose length is

$$\sqrt{A_x^2 + A_y^2}$$

is one leg in a right triangle whose other leg is $A_z \overline{z}$ and whose hypotenuse is vector $\overline{A}$. Applying the Pythagorean theorem again, we find that the length of $\overline{A}$ is

$$\sqrt{A_x^2 + A_y^2 + A_2^2}$$

Substituting our values we have

$$\sqrt{3^2 + 1^2 + 2^2} = \sqrt{14}$$

b) We refer again to Figure 1 (on previous page). The projection of $\overline{A}$ on the xy plane is simply the dotted line that is the vector

$$A_x \overline{x} + A_y \overline{y} + A_z \overline{z}$$

Its length is

$$\sqrt{A_x^2 + A_y^2}$$

by the Pythagorean theorem. In our problem, the length is

$$\sqrt{3^2 + 1^2} = \sqrt{10}$$

c) Construct a vector in the xy plane and perpendicular to A. We want a vector of the form

$$B = B_x \overline{x} + B_y \overline{y}$$

with the property $\overline{A} \bullet \overline{B} = 0$ (since

$$\overline{A} \bullet \overline{B} = |\overline{A}| \, |\overline{B}| \cos \phi$$

where ϕ is the angle between $\overline{A}$ and $\overline{B}$). Hence,

$$(3\overline{x} + \overline{y} + 2\overline{z}) \bullet (B_x \overline{x} + B_y \overline{y}) = 0$$

On taking the scalar product we find

$$3B_x + B_y = 0,$$

or

$$\frac{B_y}{B_x} = -3$$

The length of vector B is not determined by the specification of the problem. We have therefore, determined just the slope of vector B, not its magnitude. See Figure 2 (on page 13).

d) The unit vector $\overline{B}$ is the vector in the B direction but with the magnitude 1. It lies in the xy plane, and its slope (B_y / B_x) is equal to − 3. Therefore, $\overline{B}$ must satisfy the following two equations:

$$B_x^2 + B_y^2 = 1$$

$$\frac{\overline{B}_y}{\overline{B}_x} = -3$$

Solving simultaneously we have:

$$B_x^2 + (-3B_x)^2 = 1$$

or $$B_x = \frac{1}{\sqrt{10}}$$

and $$B_y = \frac{-3}{\sqrt{10}}$$

The vector $\overline{B}$ is then:

$$\overline{B} = \left(\frac{1}{\sqrt{10}}\right)\overline{x} + \left(\frac{-3}{\sqrt{10}}\right)\overline{y}$$

e) Converting the vectors into coordinate form and computing the dot product (scalar product):

$$(3\overline{x} + \overline{y} + 2\overline{z}) \bullet (2\overline{x} + 0\overline{y} + 0\overline{z}) = 6 + 0 + 0 = 6$$

f) Find the form of $\overline{A}$ and $\overline{C}$ in a reference frame obtained from the old reference frame by a rotation of $\pi/2$ clockwise looking along the positive z axis. The new unit vectors $\overline{x}, \overline{y}$, and $\overline{z}$ are related to the old $\overline{x}, \overline{y}$, and $\overline{z}$ by (see Figure 3 on page 13):

$$\overline{x}' = \overline{y}; \ \overline{y}' = \overline{x}; \overline{z}' = \overline{z}$$

Where $\overline{x}$ appeared we now have $-\overline{y}'$; where $\overline{y}$ appeared, we now have $\overline{x}'$, so that

$$\overline{A} = \overline{x} - 3\overline{y}' + 2\overline{z}' \quad \overline{c} = -2\overline{y}'$$

g) Using the results of part (f), we convert the vectors $\overline{A}$ and $\overline{C}$ into coordinate form in the primed coordinate system, giving us the following dot product:

$$\overline{A} \bullet \overline{C} = (\overline{x}' - 3\overline{y}' + 2\overline{z}') \bullet (0\overline{x}' - 2\overline{y}' + 0\overline{z}') = 0 + 6 + 0 = 6$$

This is exactly the result obtained in the unprimed system.

h) Find the vector product $\overline{A} \times \overline{C}$. In the unprimed system $\overline{A} \times \overline{C}$ is defined as

$$= (1 \bullet 0 - 2 \bullet 0)\overline{x} - (3 \bullet 0 - 2 \bullet 2)\overline{y} + (3 \bullet 0 \bullet 2.1)\overline{z}$$

$$\begin{vmatrix} \overline{x} & \overline{y} & \overline{z} \\ 3 & 1 & 2 \\ 2 & 0 & 0 \end{vmatrix} = 4_{\overline{y}} - 2_{\overline{z}}$$

i) From the vector $\overline{A} - \overline{C}$, we have

$$\overline{A} - \overline{C} = (3 - 2)\overline{x} + \overline{y} + 2\overline{z} = \overline{x} + \overline{y} + 2\overline{z}.$$

1.6   Adding Vectors Analytically

a) Analytical addition involves adding the components of the individual vectors to produce the sum, expressed in terms of its components.

b) To find $\overline{a} + \overline{b} + \overline{c}$ analytically:

   i) Resolve $\overline{a}$ in terms of its components:
$$\overline{a} = \overline{i}a_x + \overline{j}a_y$$

   ii) Resolve $\overline{b}$ in terms of its components:
$$\overline{b} = \overline{i}b_x + \overline{j}b_y$$

iii) The components of $\bar{c}$ equal the sum of the corresponding components of $\bar{a}$ and $\bar{b}$:

$$\bar{c} = \bar{i}(a_x + b_x) + \bar{j}(a_y + b_y)$$

and
$$\bar{c} = \bar{i}c_x + \bar{j}c_y$$

and the magnitude
$$|c| = \sqrt{c_x^2 + c_y^2}$$

with θ given by
$$\tan\theta = \frac{c_y}{c_x}$$

Problem Solving Example:

Q Two wires are attached to a corner fence post with the wires making an angle of 90° with each other. If each wire pulls on the post with a force of 50 kg, what is the resultant force acting on the post? See figure.

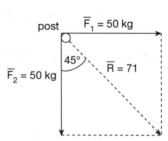

A As shown in the figure, we complete the parallelogram. If we measure $\bar{R}$ and scale it, we find it is equal to about 71 kg. The angle of the resultant is 45° from either of the component vectors.

If we use the fact that the component vectors are at right angles to each other, we can write

$$R^2 = 50^2 + 50^2$$

whence $\bar{R}$ = 71 kg approximately at 45° to each wire.

1.7 Multiplication of Vectors

a) Multiplication of a Vector by a Scalar

The product of a vector $\bar{a}$ and a scalar k, written as $k\bar{a}$, is a new vector whose magnitude is k times the magnitude of $\bar{a}$; if k is positive, the new vector has the same direction as $\bar{a}$; if k is negative, the new vector has a direction opposite that of $\bar{a}$.

b) The Scalar Product (Dot Product)

The dot product of two vectors yields a scalar:

$$\bar{a} \bullet \bar{b} = ab \cos\theta$$

c) The Vector Product (Cross Product)

i) The Cross Product of two vectors yields a vector:

$$\bar{a} \times \bar{b} = \bar{c} \quad \text{and} \quad |\bar{c}| = ab \sin\theta$$

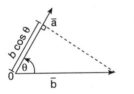 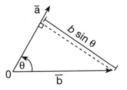

(a) Dot Product   **(b) Cross Product**

Figure 1.6 Vector Multiplication

ii) The direction of the vector product $\bar{a} \times \bar{b} = \bar{c}$ is given by the "Right-Hand Rule":

aa) With $\bar{a}$ and $\bar{b}$ tail-to-tail, draw the angle θ from $\bar{a}$ to $\bar{b}$.

bb) With your right hand, curl your finger in the direction of the angle drawn. The extended thumb points in the direction of $\bar{c}$.

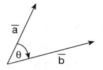

Figure 1.7 Direction of a Vector

The direction of the vector product

$$\bar{c} = \bar{a} \times \bar{b} = \quad (|\bar{c}| = ab \sin \theta)$$

is into the page.

Problem Solving Example:

Show that the area of a parallelogram, whose sides are formed by the vectors $\bar{A}$ and $\bar{B}$ (see figure), is given by

$$\text{Area} = \left| \bar{A} \times \bar{B} \right|$$

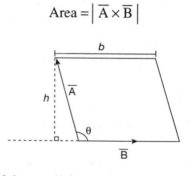

The area of the parallelogram shown in the figure is

$$\text{Area} = bh$$

But $h = |\bar{A}| \sin \theta$ and $b = |\bar{B}|$

$$\text{Area} = |\bar{A}| \, |\bar{B}| \sin \theta$$

The right side of this equation is the magnitude of $\overline{A} \times \overline{B}$, hence

$$\text{Area} = |\,\overline{A} \times \overline{B}\,| = |\,\overline{A}\,| \cdot |\,\overline{B}\,| \sin \theta$$

If we are interested in obtaining a vector area, we may write

$$\text{Area} = \overline{A} \times \overline{B}$$

where the direction of the area is the direction of $\overline{A} \times \overline{B}$. Such vector areas are useful in defining certain surface integrals used in physics.

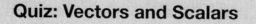

Quiz: Vectors and Scalars

1. According to the diagram, the x component of A is

   (A) A tan 20°.

   (B) A cos 20°.

   (C) A sin 20°.

   (D) 0.

   (E) A.

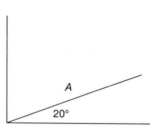

QUESTIONS 2 and 3 refer to the following diagram, passage, and choices. An object is fired horizontally from the top of a building A and follows a free-fall trajectory as shown (to B). Neglecting air friction, consider the five vectors shown.

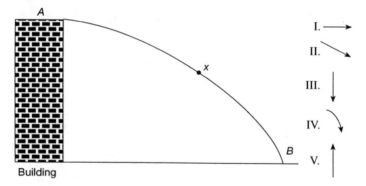

2. Which vector best represents the direction of the object's velocity at x?

   (A) I   (D) IV

   (B) II   (E) V

   (C) III

3. Which vector best represents the direction of the object's acceleration at B?

(A) I

(D) IV

(B) II

(E) V

(C) III

QUESTIONS 4–7 relate to the following properties and diagram of a ball being twirled vertically by a string.

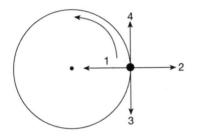

I. Centripetal force on the ball

II. Centrifugal force on the ball

III. Weight of the ball

IV. Path of the ball

V. Velocity of the ball

4. Which property is described by vector #1?

(A) I

(D) IV

(B) II

(E) V

(C) III

5. Which property is described by vector # 2?

    (A) I   (D) IV

    (B) II   (E) V

    (C) III

6. Which property is described by vector # 3?

    (A) I   (D) IV

    (B) II   (E) V

    (C) III

7. Which property is described by vector # 4?

    (A) I   (D) IV

    (B) II   (E) V

    (C) III

QUESTIONS 8 and 9 refer to the following five vector diagrams.

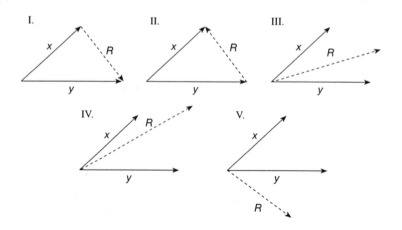

8. Which vector diagram correctly demonstrates $x + y$?

   (A) I   (D) IV

   (B) II   (E) V

   (C) III

9. Which vector diagram correctly demonstrates $x - y$?

   (A) I   (D) IV

   (B) II   (E) V

   (C) III

10. Projectile B is launched at an angle of 45° with the horizontal and lands at point x. Where did projectiles A and C land if they were launched with the same velocity as projectile B?

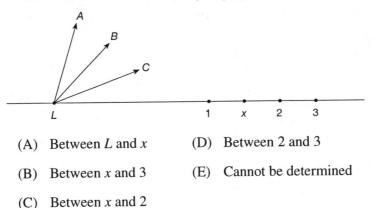

   (A) Between L and x (D) Between 2 and 3

   (B) Between x and 3   (E) Cannot be determined

   (C) Between x and 2

ANSWER KEY

1. (B)

2. (B)

3. (C)

4. (A)

5. (B)

6. (C)

7. (E)

8. (C)

9. (B)

10. (A)

CHAPTER 2

One-dimensional Motion

2.1 Basic Definitions

a) Average Velocity

$$v_{\text{avg}} = \frac{\Delta x}{\Delta t} = \frac{x_2 - x_1}{t_2 - t_1} \qquad \text{units:} \quad \frac{\text{meters}}{\text{sec}}$$

Figure 2.1 Average Velocity is a Function of Distance Over Time.

b) Instantaneous Velocity

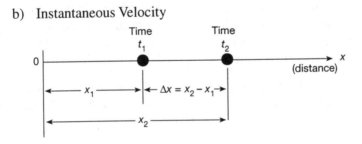

$$v = \lim_{\Delta t \to 0} \frac{\Delta x}{\Delta t} = \frac{dx}{dt} = v(t) \qquad \text{units:} \quad \frac{\text{meters}}{\text{sec}}$$

c) Average Acceleration

$$a = \frac{\Delta v}{\Delta t} = \frac{v_2 - v_1}{t_2 - t_1} \quad \text{units:} \quad \frac{\text{meters / sec}}{\text{sec}} = \frac{\text{meters}}{\text{sec}^2}$$

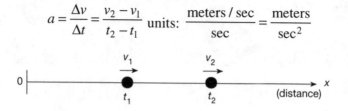

Figure 2.2 Average Acceleration is a Function of Velocity Over Time.

v_1 = initial velocity

v_2 = final velocity

d) Instantaneous Acceleration

$$a = \lim_{\Delta_t \to 0} \frac{\Delta v}{\Delta t} = \frac{dv}{dt} = a(t) \quad \text{units:} \quad \frac{\text{meters}}{\text{sec}^2}$$

Problem Solving Examples:

 An automobile driver, A, traveling relative to the earth at 65 mi/hr on a straight, level road, is ahead of motorcycle officer B, traveling in the same direction at 80 mi/hr. What is the velocity of B relative to A? Find the same quantity if B is ahead of A.

 The velocity of B relative to A is equal to the velocity of B relative to the earth minus the velocity of A relative to the earth,

or

$$V_{BA} = V_{BE} - V_{AE}$$
$$= 80 \text{ mi/hr} - 65 \text{ mi/hr}$$
$$= 15 \text{ mi/hr}$$

If B is ahead of A, the velocity of B relative to A is still the velocity of B relative to the earth minus the velocity of A relative to the earth or 15 mi/hr.

In the first case, B is overtaking A, and, in the second, B is pulling ahead of A.

Q At $t_1 = 0$ an automobile is moving eastward with a velocity of 30 km/hr. At $t_2 = 1$ min the automobile is moving northward at the same velocity. What average acceleration has the automobile experienced?

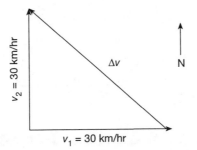

A Since velocity is a vector quantity, vector addition must be used to solve this problem. Geometrically, when two vectors are added, the tail of the second vector is placed at the head of the first and the resultant vector is drawn from the tail of the first to the head of the second. To find the difference between the two velocities, we write

$$\bar{v}_2 - \bar{v}_1 = \Delta\bar{v}$$

Changing the expression above into one including only addition:

$$\bar{v}_2 = \Delta\bar{v} + \bar{v}_1$$

This is shown in the accompanying vector diagram.

The magnitude of Δv is (refer to the figure and use the Pythagorean theorem):

$$\Delta v = \sqrt{(30 \text{ km/hr})^2 + (30 \text{ km/hr})^2}$$

$$= \sqrt{1,800 \text{ (km/hr)}^2}$$

$$= 42.4 \text{ km/hr}$$

The magnitude of the average acceleration is

$$\bar{a} = \frac{\Delta v}{\Delta t}$$

$$= \frac{42.4 \text{ km/hr}}{\dfrac{1}{60} \text{ hr}} = 2,544 \frac{\text{km/hr}}{\text{hr}}$$

The direction of Δv and hence the direction of a is, from the figure, in the direction northwest.

2.2 Motion with Constant Acceleration

$$a = \text{constant}$$

a) The velocity,

$$v \left(\frac{\text{meters}}{\text{sec}} \right)$$

i) In terms of a and t,

$$v = v_0 + at$$

where $v_0 = v(0)$ is the velocity at time $t = 0$ (the initial velocity).

ii) In terms of a and x,

$$v^2 = v_0^2 + 2a(x - x_0)$$

or

$$v = \sqrt{v_0^2 + 2a(x - x_0)}$$

where $x_0 = x(0)$ is the position at time $t = 0$, and v_0 is the velocity at time $t = 0$.

b) The position, x (meters)

i) In terms of a and t,

$$x = x_0 + v_0 t + \frac{1}{2} a t^2$$

ii) In terms of v and t,

$$x = x_0 + \frac{1}{2}(v_0 + v)t$$

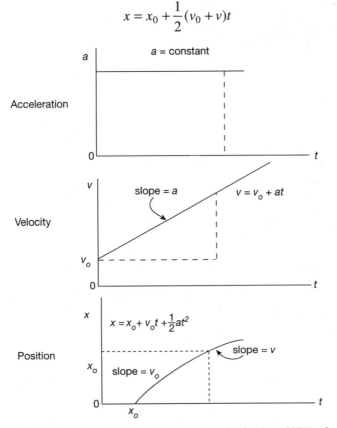

Figure 2.3 Graphs of One-dimensional Motion With Constant Acceleration.

Problem Solving Example:

Q A boy can throw a baseball horizontally with a speed of 20 m/sec. If he performs this feat in a convertible that is moving at 30 m/sec in a direction perpendicular to the direction in which he is throwing (see figure), what will be the actual speed and direction of motion of the baseball?

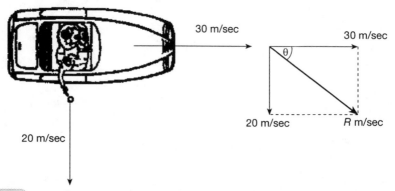

A Since the baseball is originally traveling with the convertible, it has the speed of 30 m/sec in the direction the car is traveling. When the boy throws the ball perpendicular to the car's path, he imparts an additional velocity of 20 m/sec in that direction. The ball's velocity is then 30 m/sec in the direction the convertible is moving and 20 m/sec perpendicular to this movement. Its resultant velocity can be found through adding vectors as shown in the diagram.

If the resultant velocity is R m/sec at an angle θ to the direction in which the convertible is moving, then

$$R^2 = (20)^2 + (30)^2 = 1{,}300$$

$$R = \sqrt{1{,}300} = 36.06 \text{ m/sec}$$

Also, $$\tan \theta = \frac{20}{30} = 0.666$$

From tables of tangents, $\theta = 33.69°$. Therefore, the ball has a speed of

36.06 m/sec in a direction at an angle of 33.69° to the direction in which the convertible is traveling.

2.3 Freely Falling Bodies

$$a_y = \text{constant} = -g$$

$$g = 32.2 \text{ ft/sec}^2 = 9.81 \text{ m/sec}^2$$

The equations of motions are the same as for other systems of constant acceleration. One must note, however, that a falling object has a negative velocity and a negative acceleration ($-g$), with respect to a positive height, y.

Problem Solving Example:

 A body is released from rest and falls freely. Compute its position and velocity after 1, 2, 3, and 4 seconds. Take the origin O at the elevation of the starting point, the y-axis vertical, and the upward direction as positive.

The initial coordinate y_0 and the initial velocity v_0 are both zero (see figure on the following page). The acceleration is downward, in the negative y-direction, so $a = -g = -9.8 \text{ m/sec}^2$.

Since the acceleration is constant, we may use the kinematical equations for constant acceleration, or

$$y = v_0 t + \frac{1}{2}at^2 = 0 - \frac{1}{2}gt^2 = -\frac{4.9 \text{ m}}{\text{sec}^2} \times t^2$$

$$v = v_0 + at = 0 - gt = -\frac{9.8 \text{ m}}{\text{sec}^2} \times t$$

When $t = 1$ sec,

$$y_1 = \frac{-4.9 \text{ m}}{\text{sec}^2} \times 1 \text{ sec}^2 = -4.9 \text{ m}$$

$$v_1 = \frac{-9.8 \text{ m}}{\text{sec}^2} \times 1 \text{ sec} = -\frac{9.8 \text{ m}}{\text{sec}}$$

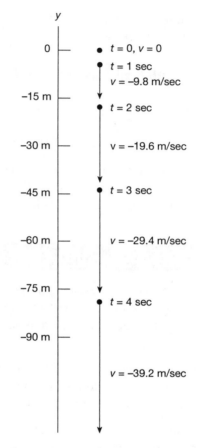

The body is therefore 4.9 m below the origin (y is negative) and has a downward velocity (v is negative) of magnitude 9.8 m/sec. The position and velocity at 2, 3, and 4 sec are found in the same way.

$$y_2 = -\frac{4.9 \text{ m}}{\text{sec}^2} \times (2 \text{ sec})^2 = -\frac{4.9 \text{ m}}{\text{sec}^2} \times 4 \text{ sec}^2 = -19.6 \text{ m}$$

$$v_2 = -\frac{9.8 \text{ m}}{\text{sec}^2} \times 2 \text{ sec} = -\frac{19.6 \text{ m}}{\text{sec}}$$

$$y_3 = -\frac{4.9\ \text{m}}{\text{sec}^2} \times (3\ \text{sec})^2 = -\frac{4.9\ \text{m}}{\text{sec}^2} \times 9\ \text{sec}^2 = -44.1\ \text{m}$$

$$v_3 = -\frac{9.8\ \text{m}}{\text{sec}^2} \times 3\ \text{sec} = -\frac{29.4\ \text{m}}{\text{sec}}$$

$$y_4 = -\frac{4.9\ \text{m}}{\text{sec}^2} \times (4\ \text{sec})^2 = -\frac{4.9\ \text{m}}{\text{sec}^2} \times 16\ \text{sec}^2 = -78.4\ \text{m}$$

$$v_4 = -\frac{9.8\ \text{m}}{\text{sec}^2} \times 4\ \text{sec} = -\frac{39.2\ \text{m}}{\text{sec}}$$

The results are illustrated in the diagram on the previous page.

Quiz: One-dimensional Motion

1. The time it takes for a plane to change its speed from 100 m/s to 500 m/s with a uniform acceleration in a distance of 1,200 m is

    (A) 1 s.   (D) 4 s.

    (B) 2 s.   (E) 5 s.

    (C) 3 s.

2. The graph shows the speed of an object as a function of time. The average speed of the object during the time interval shown is (see figure below)

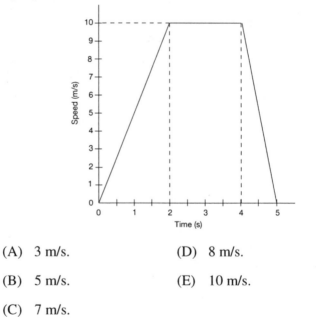

    (A) 3 m/s.   (D) 8 m/s.

    (B) 5 m/s.   (E) 10 m/s.

    (C) 7 m/s.

3. Which velocity vs. time graph best represents the speed of a falling Styrofoam ball as a function of time, taking air resistance into account?

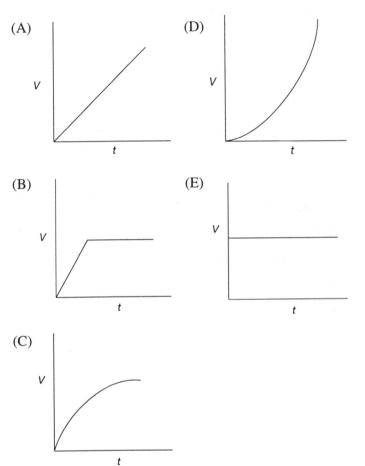

4. Uniform acceleration may be determined by any of the following methods EXCEPT

   (A) the change of velocity/time to change velocity.

   (B) speed/time.

   (C) constant force/mass.

   (D) the slope of a velocity vs. time graph.

   (E) the distance an object travels from rest/the time it travels squared.

5. An object falling in air experiences a drag force directly related to the square of the speed such that $F = Cv^2$ (where C is a constant of proportionality). Assuming that the buoyant force due to the air is negligible, the terminal velocity of this falling body is best described by the equation

   (A) $\dfrac{mg}{C}$ (D) $\sqrt{\dfrac{mg}{C}}$

   (B) $\dfrac{\sqrt{mg}}{C}$ (E) $\dfrac{C}{mg}$

   (C) $\dfrac{mg}{\sqrt{C}}$

6. Consider that a coin is dropped into a wishing well. You want to determine the depth of the well from the time T between releasing the coin and hearing it hit the bottom. Suppose $T = 2.059$ s. What is the depth h of the well (see figure on next page)?

(A) 20.77 m

(B) 19.60 m

(C) 23,564 m

(D) 18.43 m

(E) 39.20 m

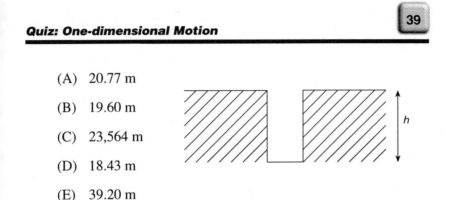

7. Consider an hourglass on a scale pictured below at times $t = 0$, 0.001, and 1 hour. What happens to the scale's measure of weight of the hourglass plus sand combination as the sand falls?

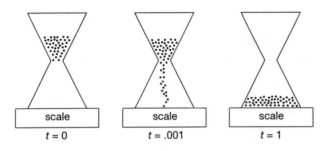

(A) The weight is constant.

(B) The weight decreases and then increases.

(C) The weight increases.

(D) The weight increases and then decreases.

(E) The weight is unchanged.

8. A graph of displacement vs. time for an object moving in a straight line is shown below. The acceleration of the object must be

(A) zero.

(B) increasing.

(C) decreasing.

(D) constant and greater than zero.

(E) equal to *g*.

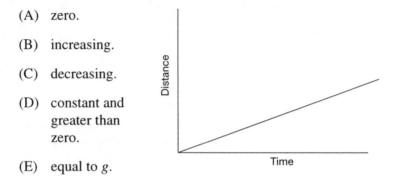

QUESTIONS 9 and 10 relate to the following graph of distance vs. time for a motion along a straight line.

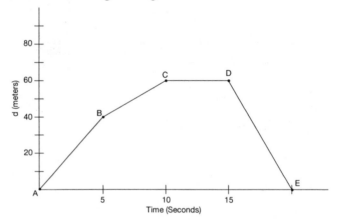

9. What is the velocity from *A* to *B*?

(A) 20 m/sec

(B) 8 m/sec

(C) 200 m/sec

(D) .125 m/sec

(E) None of the above.

10. What distance is travelled from *C* to *D*?

    (A) 60 meters   (D) 600 meters

    (B) 30 meters   (E) 0 meters

    (C) 80 meters

ANSWER KEY

1.	(D)	6.	(B)
2.	(C)	7.	(B)
3.	(B)	8.	(A)
4.	(B)	9.	(B)
5.	(D)	10.	(E)

CHAPTER 3

Plane Motion

3.1 Displacement, Velocity, and Acceleration in General Planar Motion

a) The Displacement (Position) Vector

$$\bar{r} = \bar{i}x + \bar{j}y$$

b) The Velocity Vector

$$\bar{v} = \frac{d\bar{r}}{dt} = \bar{i}v_x + \bar{j}v_y$$

c) The Acceleration Vector

$$\bar{a} = \frac{d\bar{v}}{dt} = \bar{i}a_x + \bar{j}a_y$$

The vectors $\bar{r}$, $\bar{v}$, $\bar{a}$, and their components are shown in Figure 3.1, on the next page.

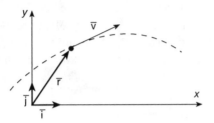

(a) A particle moves in a curved path in the *xy* plane. At time *t* the position vector is $\bar{r}$ and its velocity is $\bar{v}$.

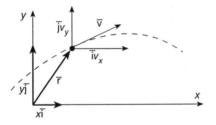

(b) The components of vectors $\bar{r}$ and $\bar{v}$, at time *t*.

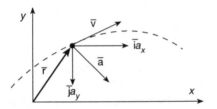

(c) The particle experiences an acceleration $\bar{a}$ at time *t*. The components are a_x and a_y. Notice that a_y is negative.

Figure 3.1 A Particle Traveling in Plane Motion

Problem Solving Example:

A man standing on the roof of a building 30 meters high throws a ball vertically downward with an initial velocity of 500 cm/

sec^{-1} as it leaves his hand (see figure). The acceleration due to gravity is 9.8 m/sec^{-2}. (a) What is the velocity of the ball after it has been falling for 0.5 seconds? (b) Where is the ball after 1.5 seconds? (c) What is the velocity of the ball as it strikes the ground?

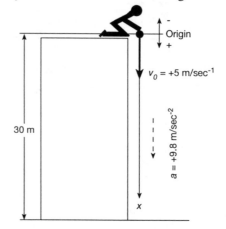

$$v = v_0 + at = (+5 \text{ m/s}) + (9.8 \text{ m/s}^2)(0.5 \text{ s})$$
$$= 5 \text{ m/sec} + 4.9 \text{ m/sec} = +9.9 \text{ m/sec}$$

A We will use the MKS system of units because most of the given quantities are expressed in these units. Then the initial velocity must be expressed as 5 meters per second since 1 cm/s = .01 m/s. Place the origin at the top of the building. Then $x = 0$ when $t = 0$. Let the positive direction of x be downward. The initial velocity is downward, and, therefore, positive, so $v_0 = +5$ m/sec^{-1}. The acceleration is downward and, therefore, positive, so $a = +9.8$ m/sec^{-2}.

a) In this part of the problem, one is given v_0, a, and t, and must deduce v. Since the acceleration is constant, we may use the kinematics equations for constant acceleration, or

$$v = v_0 + at = (+5 \text{ m/s}) + (9.8 \text{ m/s}^2)(0.5 \text{ s})$$
$$= 5 \text{ m/sec} + 4.9 \text{ m/sec} = +9.9 \text{ m/sec}$$

After 0.5 second the velocity is 9.9 m/sec^{-1} downward.

b) In this part of the problem one is given v_0, a, and t, and must calculate x. By definition of velocity,

$$v = \frac{dx}{dt}$$

Therefore,

$$\int_{x_0}^{x} dx = \int_{t=0}^{t} v\, dt,$$

where x_0, is the initial position of the ball. Using the formula for v given in the previous part, we have

$$x - x_0 = \int_{0}^{t} (v_0 + at)\, dt$$

or

$$x = x_0 + v_0 t + \frac{1}{2} at^2$$

Because $x_0 = 0$

$$x = v_0 t + \frac{1}{2} at^2$$

$$= (+5 \text{ m/sec})(+1.5 \text{ sec}) + \frac{1}{2}(9.8 \text{ m/sec}^2)(+1.5 \text{ sec})^2$$

$$= 7.5 \text{ m} + (4.9 \text{ m/sec}^2)(2.25 \text{ sec}^2)$$

$$= 7.5 \text{ m} + 11.025 \text{ m}$$

$$= 18.525 \text{ m}$$

After 1.5 seconds the ball is 18.525 meters below the roof or 11.475 meters above the ground.

c) When the body strikes the ground $x = +30$ m. So one is given v_0, a, and x, and asked to calculate v. The correct equation, then, should not contain t as a variable. The equation to be used is, since $a =$ constant,

$$v^2 = v_0^2 + 2ax$$
$$= (+5 \text{ m/sec})^2 + 2(+9.8 \text{ m/sec}^2)(+30 \text{ m})$$
$$= 25 \text{ m}^2/\text{sec}^2 + 588 \text{ m}^2/\text{sec}^2$$
$$v^2 = 613 \text{ m}^2/\text{sec}^2$$
$$v = 24.76 \text{ m/sec}$$

When it strikes the ground, the ball has a velocity of 24.76 m/sec^{-1}.

3.2 Motion in a Plane with Constant Acceleration

The Equations of Motion

a) $\bar{a} = $ constant

b) $a_x = $ constant

 $a_y = $ constant

The Velocity Vector, $\bar{v}$

c) $\bar{v} = \bar{i}v_x + \bar{j}v_y$

i) The components of velocity in the x- and y-directions are

$$v_x = v_{x_0} + a_x t$$

and $$v_y = v_{y_0} + a_y t$$

ii) In terms of position and acceleration,

$$v_x^2 = v_{x_0}^2 + 2a_x (x - x_0)$$

and
$$v_y^2 = v_{y_0}^2 + 2a_y (y - y_0)$$

The Velocity Vector, $\bar{v}$

$$\bar{v} = \bar{v}_0 + \bar{a}t$$

where
$$\bar{v}_0 = \bar{i}v_{0_x} + \bar{j}v_{0_y}$$

(the initial velocity vector) and

$$\bar{a} = \bar{i}a_x + \bar{j}a_y$$

The Position Vector, $\bar{r}$

$$\bar{r} = x\bar{i} + y\bar{j}$$

where
$$x = x_0 + \frac{1}{2}(v_{x_0} + v_x)t$$

or
$$x = x_0 + v_{x_0}t + \frac{1}{2}a_x t^2$$

and
$$y = y_0 + \frac{1}{2}(v_{y_0} + v_y)t$$

or
$$y = y_0 + v_{y_0}t + \frac{1}{2}a_y t^2$$

The position vector may also be written as

$$\bar{r} = \bar{r}_0 + \bar{v}_0 t + \frac{1}{2}\bar{a}t^2$$

Problem Solving Example:

The pilot of an airplane flying on a straight course knows from his instruments that his airspeed is 300 kph. He also knows

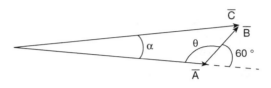

that a 60-kph gale is blowing at an angle of 60° to his course. How can he calculate his velocity relative to the ground?

A Relative to an observer on the ground, the airplane has two velocities, one of 300 kph relative to the air and the other 60 kph at an angle of 60° to the course, due to the fact that it is carried along by the moving air mass.

To obtain the resultant velocity, it is therefore necessary to add the two components by vector addition. In the diagram, $\overline{A}$ represents the velocity of the aircraft relative to the air, and $\overline{B}$ the velocity of the air relative to the ground. When they are added in the normal manner of vector addition, $\overline{C}$ is their resultant. The magnitude of $\overline{C}$ is given by the trigonometric formula known as the law of cosines (see figure).

$$C^2 = A^2 + B^2 - 2AB \cos\theta$$

But $A = 300$ kph, $B = 60$ kph, and $\theta = (180° - 60°) = 120°$. Therefore,

$$C^2 = (300 \text{ kph})^2 + (60 \text{ kph})^2 - 2 \times 300 \text{ kph} \times 60 \text{ kph} - \frac{(1)}{2}$$

$$= 111,600 \text{ kph}^2$$

$$\therefore C = 334 \text{ kph}$$

Also, from the addition formula for vectors, we have

$$\sin\alpha = \frac{B}{C}\sin\theta = \frac{60 \text{ kph}}{334 \text{ kph}} \times \frac{\sqrt{3}}{2} = 0.156$$

$$\therefore \alpha = 9°$$

3.3   Projectile Motion

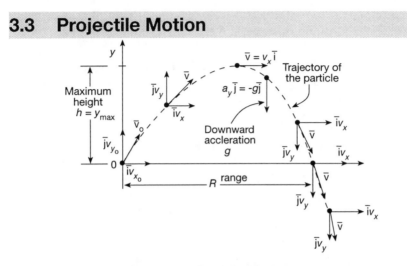

Figure 3.2 Particle Motion

Acceleration in constant

$$a_x = 0$$
$$a_y = -g$$
$$g = 9.81 \text{ m/sec}^2 = 32.3 \text{ ft/sec}^2$$

The initial conditions:

$$x_0 = 0$$
$$y_0 = 0$$
$$\overline{v}_0 = \overline{i}v_{x_0} + \overline{j}v_{y_0}$$
$$v_{x_0} = v_0 \cos\theta_0$$
$$v_{y_0} = v_0 \sin\theta_0$$

a)   The Velocity

The magnitude of velocity at any instant t is

$$v = \sqrt{v_x^2 + v_y^2}$$

The horizontal component of velocity is constant throughout the projectile motion since there is no horizontal acceleration:

$$v_{x_0} = v_0 \cos\theta_0 = \text{constant}$$

The vertical component of velocity is

$$v_y = v_{y_0} - gt$$

or $$v_y = (v_0 \sin\theta_0) - gt$$

since the negative acceleration of gravity acts vertically.

b) The Position

The Horizontal Position Component is

$$x = v_x t$$

or $$x = (v_0 \cos\theta_0)t$$

The Vertical Position Component is

$$y = v_{y_0} t - \frac{1}{2}gt^2$$

or $$y = (v_0 \sin\theta_0)t - \frac{1}{2}gt^2$$

or as a function of the horizontal position component,

$$y = (\tan\theta_0)x - \frac{g}{2(v_0 \cos\theta_0)}x^2$$

This is the trajectory of the particle. The angle made by the velocity vector with the horizontal at any instant is

$$\tan\theta = \frac{v_y}{v_x}$$

or $\qquad \theta = \tan^{-1}\frac{v_y}{v_x}$

i) The Range

In terms of v_0 and θ_0,

$$R = \frac{v_0^2}{g}\sin 2\theta_0$$

In terms of time,

$$R = v_{x_0}t_2 = (v_0\cos\theta_0)t_2$$

Here, t_2 is the time required for the particle to traverse the full range. The maximum range is obtained when the initial angle of flight is

$$\theta_0 = 45° = \frac{\pi}{4}\ \text{rad}$$

$$R_{\max} = \frac{v_0^2}{g}$$

ii) The Maximum Height (Elevation)

In terms of v_0 and θ_0,

$$y_{\max} = \frac{v_0^2\sin^2\theta_0}{2g} = \frac{(v_{y_0})^2}{2g}$$

The time t_1 required to attain maximum height is

$$t_1 = \frac{v_{y_0}}{g} = \frac{1}{2}t_2$$

Problem Solving Example:

 A ball is projected horizontally with a velocity v_0 of 3 m/sec. Find its position and velocity after $^1/_4$ sec (see the figure).

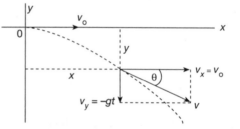

Since the acceleration of gravity, g, is constant, we may use the equations for constant acceleration to find the velocity (v_y) and position (y) of a particle undergoing free-fall motion:

$$v_y = v_{y_0} - gt$$

$$y = y_0 + v_{y_0} - \frac{1}{2}gt^2$$

Here, y_0 and v_{y_0} are the initial y position and velocity of the particle. In this case, the departure angle is zero. The initial vertical velocity component is therefore zero. The horizontal velocity component equals the initial velocity and is constant. Since no horizontal force sets on the flying object, it is not accelerated in the horizontal direction. Therefore,

$$v_y = -gt \qquad\qquad y = -\frac{1}{2}gt^2$$

$$v_x = v_{x_0} \qquad\qquad x = v_{x_0}t$$

and, at $t = \dfrac{1}{4}\,\text{sec},$

$$y = \left(-\dfrac{1}{2}\right)(9.8\ \text{m/sec}^2)\left(\dfrac{1}{16}\ \text{sec}^2\right) = 0.30\ \text{m}$$

$$x = (2.54\ \text{m/sec})\left(\dfrac{1}{4}\ \text{sec}\right) = 0.61\ \text{m}$$

$$v_y = (-9.8\ \text{m/sec}^2)\left(\dfrac{1}{4}\ \text{sec}\right)$$

$$= -2.45\ \text{m/sec}$$

$$v_x = -2.45\ \text{m/sec}$$

3.4 Uniform Circular Motion

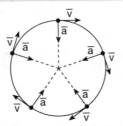

Figure 3.3 A Particle in Uniform Circular Motion

Notice that the acceleration vector is always directed toward the center of the circle and thus perpendicular to $\overline{v}$.

The magnitude of velocity is constant.

$$v = \text{constant}$$

a) Centripetal Acceleration

$$a = \dfrac{v^2}{r}$$

where:

a = acceleration

v = tangential component of velocity

r = radius of the path

b) For uniform circular motion, a can also be written as

$$a = \frac{4\pi^2 r}{T^2}$$

where T, the period or time for one revolution, is given by

$$T = \frac{2\pi r}{v}$$

c) The tangential component of the acceleration is the rate at which the particle speed changes:

$$a_T = \lim_{\Delta t \to 0} \frac{\Delta v}{\Delta t} = \frac{dv}{dt}$$

Problem Solving Example:

Q The moon revolves about the earth in a circle (very nearly) of radius $R = 358{,}500$ km or 3.6×10^8 m, and requires 27.3 days or 23.4×10^5 sec to make a complete revolution. (a) What is the acceleration of the moon toward the earth? (b) If the gravitational force exerted on a body by the earth is inversely proportional to the square of the distance from the earth's center, the acceleration produced by this force should vary in the same way. Therefore, if the acceleration of the moon is caused by the gravitational attraction of the earth, the ratio of the moon's acceleration to that of the falling body at the earth's surface should equal the ratio of the square of the earth's radius (5,925 km or 5.9×10^6 m) to the square of the radius of the moon's orbit. Is this true?

a) The velocity of the moon is

$$v = \frac{\text{distance}}{\text{time}} = \frac{\text{circumference}}{\text{time for one orbit}}$$

$$= \frac{2\pi R}{T} = \frac{2\pi \times 3.6 \times 10^8 \, \text{m}}{23.4 \times 10^5 \, \text{sec}}$$

$$= 967 \, \text{m/sec}$$

Its radial acceleration is therefore,

$$a = \frac{V^2}{R} = \frac{(967 \, \text{m/sec})^2}{3.6 \times 10^8 \, \text{m}}$$

$$= 0.0026 \, \text{m/sec}^2 = 2.6 \times 10^{-3} \, \text{m/sec}^2$$

b) The ratio of the moon's acceleration to the acceleration of a falling body at the earth's surface is

$$\frac{a}{g} = \frac{2.60 \times 10^{-3} \, \text{m/sec}^2}{9.8 \, \text{m/sec}^2}$$

$$= 3.06 \times 10^{-4}$$

The ratio of the square of the earth's radius to the square of the moon's orbit is

$$\frac{(5.9 \times 10^6 \, \text{m})^2}{(3.6 \times 10^8 \, \text{m})^2} = 3.06 \times 10^{-4}$$

The agreement is very close, although not exact because we have used average values.

Quiz: Plane Motion

1. A graph of displacement vs. time for an object moving in a straight line is shown below. The acceleration of the object must be

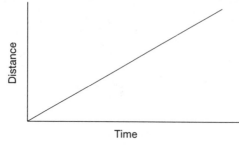

(A) zero.

(B) increasing.

(C) decreasing.

(D) constant and greater than zero.

(E) equal to g.

2. The graph on the next page shows the speed of an object as a function of time. The average speed of the object during the time interval shown is

(A) 3 m/s.   (D) 8 m/s.

(B) 5 m/s.   (E) 10 m/s.

(C) 7 m/s.

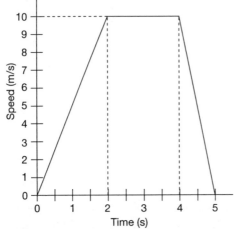

3. Projectile B is launched at an angle of 45° with the horizontal and lands at point x. Where did projectiles A and C land if they were launched with the same velocity as projectile B?

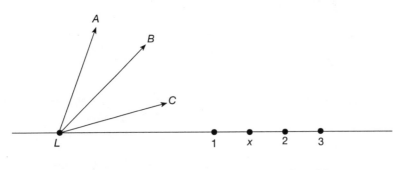

(A) Between L and x

(D) Between 2 and 3

(B) Between x and 3

(E) Cannot be determined.

(C) Between x and 2

4. A ball is thrown upwards from the ground and follows the path shown in the diagram on the next page. At which point does the ball have the greatest speed?

(A) Point *A*

(B) Point *B*

(C) Point *C*

(D) Points *A* and *B*

(E) Points *A* and *C*

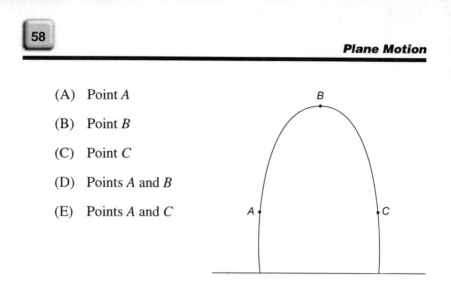

5.   A ball is thrown horizontally from the top of a tower 40 m high.
     The ball strikes the ground at a point 80 m from the bottom of the
     tower. Find the angle that the velocity vector makes with the hori-
     zontal just before the ball hits the ground.

(A) 315°   (D) 90°

(B) 41°   (E) 82°

(C) 0°

6.   Suppose that a man jumps off a building 202 m high onto cush-
     ions having a total thickness of 2 m. If the cushions are crushed to
     a thickness of 0.5 m, what is the man's acceleration as he slows
     down?

(A) 1g

(B) 133 g

(C) 5 g

(D) 2 g

(E) 266 g

7. As the ball rolls down the hill as shown, its

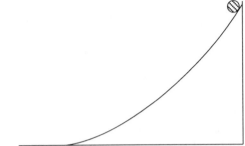

(A) speed increases and acceleration decreases.

(B) speed decreases and acceleration increases.

(C) speed increases and acceleration increases.

(D) speed decreases and acceleration decreases.

(E) Both remain constant.

QUESTIONS 8 and 9 refer to the following figure.

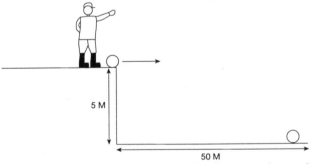

8. A child throws a baseball horizontally from an elevated position 5 meters above the ground as shown. The baseball travels a horizontal distance of 50 meters. How much time passes before the baseball hits the ground?

(A) 1 second

(B) 2 seconds

(C) 5 seconds

(D) 0.5 or $^1/_2$ second

(E) 10 seconds

9. How fast did the child throw the ball horizontally?

   (A) 5 meters/sec   (D) 500 meters/sec

   (B) 50 meters/sec   (E) 1,000 meters/sec

   (C) 100 meters/sec

10. Two balls are thrown vertically upward at the same time. Suppose that the balls have initial velocities $v_1 = 20$ m/s and $v_2 = 24$ m/s, respectively. Find the distance between the two balls when ball one is at its maximum height.

    (A) 20.40 m   (D) 8.14 m

    (B) 28.56 m   (E) 14.28 m

    (C) 16.28 m

ANSWER KEY

1.	(A)	6.	(B)
2.	(C)	7.	(A)
3.	(A)	8.	(A)
4.	(E)	9.	(B)
5.	(A)	10.	(D)

CHAPTER 4

Dynamics of a Particle

$$\overline{F} = m\overline{a} \quad \rightarrow \text{units:} \quad \left(\text{kg} \times \text{m/sec}^2\right) = \text{newtons}$$

When a body of mass m is acted upon by a force $\overline{F}$, the force $\overline{F}$ and the acceleration of the body are related by the above equation.

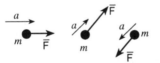

Figure 4.1 The Acceleration is in the Direction of the Applied Force

For several forces:

$$\sum \overline{F} = m\overline{a}$$

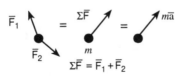

Figure 4.2 Newton's Second Law Also Holds for Several Applied Forces

Problem Solving Examples:

 What is the resultant force on a body of mass 48 kg when its acceleration is 6 m/sec²?

 The relationship between a body's acceleration and the net force on it is given by Newton's second law. The mass of the body is given; hence, the net force on the body is

$$\Sigma \overline{F} = m\overline{a} = 48 \text{ kg} \times 6 \text{ m/sec}^2 = 288 \text{ newtons}$$

 A force of 0.20 newton acts on a mass of 100 grams. What is the acceleration?

 From Newton's second law, we have

$$\overline{F} = m\overline{a}$$

$$\overline{a} = \frac{\overline{F}}{m}$$

Also, 100 grams = 0.10 kg. Therefore,

$$a = \frac{0.20 \,\text{N}}{0.10 \,\text{kg}} = \frac{0.20 \,\text{kg-m}/g^2}{0.10 \,\text{kg}}$$

4.2   Newton's Third Law

To every action there is always an equal, opposing reaction.

If a body *A* exerts a force *F* on body *B*, then body *B* exerts an equal and opposite force – *F* on body *A*.

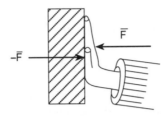

a) A wall pushes you with the same force with which you push it.

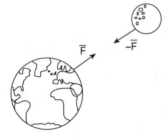

b) The gravitational force that the earth exerts on the moon is equal to and opposite to the force that the moon exerts on the earth.

c) The electrostatic force that the $+q$ charge exerts on the $-Q$ charge is equal to and opposite to the force that the $-Q$ charge exerts on the $+q$ charge.

Figure 4.3 Examples of Newton's Third Law

4.3 Mass and Weight

a) Mass → units: (kilograms (kg))

For a given body, the ratio of the magnitude of the force to that of the acceleration is a constant and is called its mass:

$$m = \frac{F}{a} = \text{constant (for a given body)}$$

b) Weight → units: (newtons)

The weight of a body is the gravitational force exerted on the body by the earth and is given by the product of the mass and the gravitational acceleration:

$$W = mg$$

Problem Solving Example:

Q A swimmer whose mass is 60 kg dives from a 3-m high platform. What is the acceleration of the earth as the swimmer falls toward the water? The earth's mass is approximately 6×10^{24} kg.

A The diver's mass, $m_d = 60$ kg, the acceleration of the diver, $a_d = 9.8$ m/s^2, and the mass of the earth, $m_e = 6 \times 10^{24}$ kg, are the known observables. The earth's acceleration can be determined using the fact the force of the earth on the diver is equal in magnitude and opposite in direction to the force of the diver on the earth. Letting the subscripts d and e refer to the diver and earth, respectively, we obtain

$$m_d \bar{a}_d = -m_e \bar{a}_e$$

Considering just the magnitude of the two vectors, we find

$$(60 \text{ kg})(9.8 \text{ m/s}^2) = (6 \times 10^{24} \text{ kg}) a_e$$

$$a_e = \frac{(60 \text{ kg})(9.8 \text{ m/s}^2)}{6 \times 10^{24} \text{ kg}}$$

$$= 9.8 \times 10^{-23} \text{ m/s}^2$$

Since the diver is accelerated downward, the earth is accelerated upward, toward the falling diver.

4.4   Friction

For impending motion,

$$\text{Frictional Force} = F_s = \mu_s N$$

μ_s = coefficient of static friction

N = normal force

For a body already in motion,

$$F_k = \mu_k N$$

μ_k = coefficient of kinetic friction

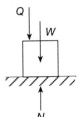

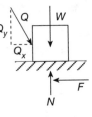

a) $Q_x = 0$, No friction

N = normal force = $Q + W$

F = friction force

= zero

b) No motion

$F = -Q_x$

$F < \mu_s N$

$N = Q_y + W$

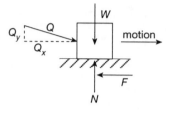
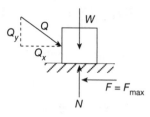

c)   Motion

$$Q_x > -F_k$$

$$F_k = \mu_k N$$

$$N = Q_y + W$$

d)   Motion impending

$$F = F_{max} = \mu_s N$$

$$-Q = F$$

$$N = Q_y + W$$

Figure 4.4   Simple Cases Involving Friction

Problem Solving Example:

Q   A 65-kg horizontal force is sufficient to draw a 1,200-kg sled on level, well-packed snow at uniform speed. What is the value of the coefficient of friction?

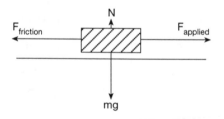

A If the sled moves at constant velocity, it experiences no net force. Therefore, the applied force must be equal to the frictional force.

$$F_{applied} = 65 \text{ kg} = F_{friction}$$

Since the frictional force is proportional to the normal force,

$$F_{friction} = \mu_k N$$

Applying Newton's second law, $\overline{F} = m\overline{a}$, to the vertical forces acting on the block, we find

$$N - mg = m\overline{a}_y$$

where $\overline{a}_y$ is the vertical acceleration. In this problem, $\overline{a}_y = 0$ because the sled doesn't rise off the surface upon which it slides. Hence,

$$N = mg$$

and $$F_{friction} = \mu_k N = \mu_k \, mg$$

But $$F_{friction} = 65 \text{ kg}$$

Therefore, $$65 = \mu_k \, mg$$

$$\mu_k = \frac{65}{1,200 \text{ kg}} = .054$$

4.5 The Dynamics of Uniform Circular Motion—Centripetal Force

The force acting on a body of mass m undergoing uniform circular motion is the centripetal force, given by

$$\overline{F} = m\overline{a} = m\frac{v^2}{r}$$

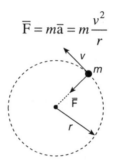

A body of mass m traveling in uniform circular motion. The body is pulled toward the center of the circle of radius r with a force

$$\overline{F} = \frac{mv^2}{r}$$

Figure 4.5 Centripetal Force of a Particle

Here, v is the magnitude of velocity, which is constant, and r is the radius of the circle.

Problem Solving Example:

 What would be the period of rotation of the earth about its axis if its rotation speed increased until an object at the equator became weightless?

 The two forces acting on a body at the equator are the force asserted on it due to the gravitational attraction of the earth, $m\,\overline{g}_0$, where $\overline{g}_0$ is the free-fall acceleration at the equator and acts toward the center, and the normal force exerted by the surface of the earth on the body, N. This latter force acts upward. On a non-rotating earth, or at the poles (which by definition do not rotate), these forces are equal since a body would be in equilibrium. At the equator, where a body does experience rotation and therefore a centripetal acceleration, the forces are unequal so that their resultant provides the centripetal force necessary to keep the body traveling in a circle. Therefore, Newton's second law yields:

$$mg_0 - N = \frac{mv^2}{R}$$

where v is the speed of the body and R is the radius of the earth. But the distance traveled in one period of rotation, T, is $2\pi R$. Therefore,

$$T = \frac{2\pi R}{v}, \text{ and } v^2 = \frac{4\pi^2 R^2}{T^2}.$$

Substituting this into the first expression, we get

$$N = mg_0 - \frac{mv^2}{R} = mg_0 - \frac{4\pi^2 mR}{T^2}$$

$$= m\left(g_0 - \frac{4\pi^2 R}{T^2}\right) = mg$$

where g is the acceleration as measured at the earth's surface, and N is a measure of the apparent weight of the body, which is thus less than the gravitational force exerted on the body by the earth. If the speed of revolution of the earth increases, the body becomes weightless when the normal force exerted on it by the surface becomes zero. Thus, weightlessness occurs when

$$N = 0 = m\left(g_0 - \frac{4\pi^2 R}{T^2}\right)$$

$$g_0 = \frac{4\pi^2 R}{T^2}$$

or when the period of rotation is

$$T = 2\pi\sqrt{\frac{R}{g_0}} = 2\pi\sqrt{\frac{4\times10^3\,\text{mi}\times5,280\,\text{ft/mi}}{32.4\,\text{ft/s}^2}}$$

$$= \frac{2\pi}{3,600\,\text{s/hr}}\times\sqrt{\frac{4\times5,280\times10^3}{32.4}}\,s = 1.41\,\text{hr}$$

Quiz: Dynamics of a Particle

1. Newton's first law of motion is often referred to as the law of

    (A) rest.   (D) inertia.

    (B) uniform velocity.   (E) apples.

    (C) centripetal acceleration.

2. Newton's second law of motion relates the acceleration of a mass
    being acted upon by an unbalanced force as being

    (A) inversely proportional to the mass.

    (B) inversely proportional to the force.

    (C) zero.

    (D) independent of the mass.

    (E) directly proportional to the force.

3. A block of weight W is pulled along a horizontal surface at a
    constant speed V by a force F, which acts at an angle θ with the
    horizontal, as shown below. The normal force exerted on the block
    by the surface is equal to

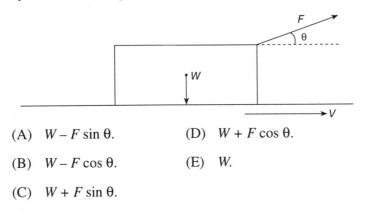

    (A) $W - F \sin \theta$.   (D) $W + F \cos \theta$.

    (B) $W - F \cos \theta$.   (E) W.

    (C) $W + F \sin \theta$.

4. If an object is moving in circular motion due to centripetal force, F, and the radius of its circular motion is then doubled, the new force then becomes

(A) $2F$.

(D) F^2.

(B) F.

(E) $\dfrac{1}{F}$.

(C) $\dfrac{F}{2}$.

5. The force needed to allow the 100 N block to go down the incline at constant speed is

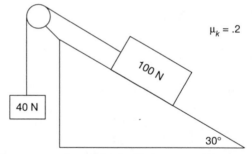

I. 7.3 newtons down the incline.

II. greater than 7.3 newtons down the incline.

III. between 7.3 newtons down and 27.3 newtons up the incline.

IV. 27.3 newtons up the incline.

V. greater than 27.3 newtons up the incline.

(A) I

(D) IV

(B) II

(E) V

(C) III

QUESTIONS 6 and 7 refer to the following information: Mass *M* slides down the incline shown with a constant velocity. Consider the following for the system.

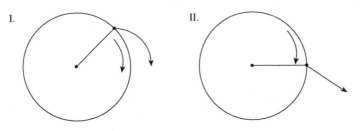

I. $\tan \theta$

II. Mg

III. $Mg \sin \theta$

IV. $Mg \cos \theta$

V. $Mg \tan \theta$

6. Which relationship equals the coefficient of sliding friction?

(A) I   (D) IV

(B) II   (E) V

(C) III

7. Which relationship equals the weight of mass *M*?

(A) I   (D) IV

(B) II   (E) V

(C) III

8. This question refers to the following diagrams.

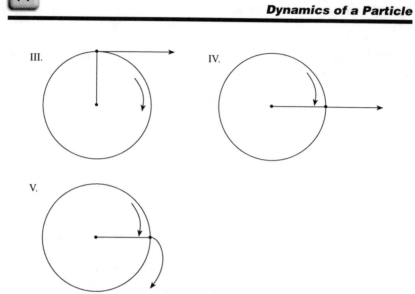

III.   IV.

V.

Imagine that a mass attached to a string is twirling clockwise in a vertical circle. When the ball reaches the points shown in the diagrams on the previous page, the string breaks. Which diagram shows the correct path followed by the mass when the string breaks.

(A) I   (D) IV

(B) II   (E) V

(C) III

QUESTIONS 9 and 10 refer to the following figure.

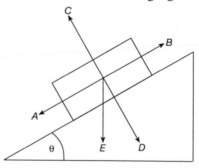

Select the letter that shows the direction of the force described in Questions 9 and 10. The block is moving down the plane.

I.	A	IV.	D
II.	B	V.	E
III	C		

9. The frictional force

   (A) I   (D) IV

   (B) II   (E) V

   (C) III

10. The force exerted by the surface

    (A) I   (D) IV

    (B) II   (E) V

    (C) III

ANSWER KEY

1.	(D)	6.	(A)
2.	(A)	7.	(B)
3.	(A)	8.	(C)
4.	(C)	9.	(B)
5.	(A)	10.	(C)

CHAPTER 5

Work and Energy

5.1 Work Done by a Constant Force

$\overline{F}$ = constant acting on a body at an angle θ in the direction of motion

d = displacement of a body

The work done by the force $\overline{F}$ on the body is

$$W = (F \cos\theta)d$$

where F = magnitude of force $\overline{F}$.

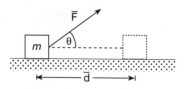

Figure 5.1 Work Done on a Block

The work may also be written as:

$$\underset{x}{\underline{W = \overline{F} \bullet \overline{d}}} \quad \text{units:} \quad \left(\frac{\text{kg} \times \text{m}}{\text{sec}^2}\right) = \text{joules}$$

where $\overline{d}$ = displacement vector

Note: Work can only be done by the component of force in the direction of displacement.

Problem Solving Examples:

Q   A suitcase is dragged 30 m along a floor by a force $F = 10$ newtons inclined at an angle 30° to the floor. How much work is done on the suitcase?

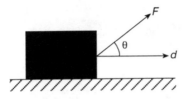

A   Work is defined as the scalar product of the force acting on an object, and the distance through which the object moves while the force is being applied.

$$W = \overline{F} \bullet \overline{d}$$

where $\overline{F}$ is the force and $\overline{d}$ is the distance. (See the figure above.) Note that the force and distance are vectors while work is a scalar, hence, the "scalar product" nomenclature for the dot.

$$W = \overline{F} \bullet \overline{d} = Fd \cos\theta = Fd \cos 30°$$

$$= (10\,\text{N})(30\,\text{m})\left(\frac{\sqrt{3}}{2}\right) = 150\,\sqrt{3}\ \text{N–m}$$

Q   A horizontal force of 5 N is required to maintain a velocity of 2 m/sec for a box of mass 10 kg sliding over a certain rough surface. How much work is done by the force in 1 minute?

First, we must calculate the distance traveled:

$$s = vt$$
$$= (2 \text{ m/sec}) \times (60 \text{ sec})$$
$$= 120 \text{ m}$$

Then, $W = Fs \cos\theta$, where θ is the angle between the force and the distance. In this case $\theta = 0°$ so we can write

$$W = Fs$$
$$= (5 \text{ N}) \times (120 \text{ m})$$
$$= 600 \text{ N–m} = 600 \text{ J}$$

5.2   Block on an Inclined Plane

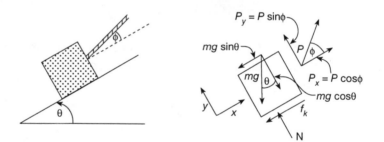

Figure 5.2 Free-body Diagram for Motion Up the Plane

a) If motion up the plane is impending,

$$\sum F_y = ma_y = P\sin\phi - mg\cos\theta + N = 0$$

b) If motion is up the plane, (f_s stands for the force of static function)

$$\sum F_x = ma_x = P\cos\phi - mg\sin\theta - f_{s,\text{maximum}} = 0$$

c) If motion is down the plane, (f_r stands for the force of kinetic friction)

$$\sum F_x = ma_x = P\cos\phi - mg\sin\theta - f_k$$

d) Work done by $\overline{P}$ to move block a distance d up the plane at constant speed (zero acceleration):

$$\sum F_x = ma_x = P\cos\phi - mg\sin\theta + f_k$$

Since we assume

$$a_x = 0, \ P\cos\phi - mg\sin\theta - f_k = 0,$$

$$W = \overline{F} \bullet \overline{d} = (P\cos\phi)\bullet d = (mg\sin\theta + f_k)\bullet d$$

Problem Solving Example:

 A 5-kg block slides down a frictionless plane inclined at an angle of 30° with the horizontal as shown in Figure (a). Calculate the amount of work W done by the force of gravity for a displacement of 8 m along the plane.

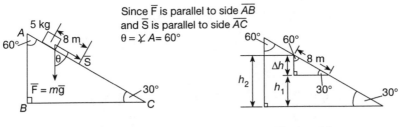

Since $\overline{F}$ is parallel to side $\overline{AB}$
and $\overline{S}$ is parallel to side $\overline{AC}$
$\theta = \angle A = 60°$

Figure (a)   Figure (b)

 We will solve this problem first by the dynamics method and then by the energy method.

The formula for calculating work is

$$W = \overline{F} \bullet \overline{s}$$

$$= Fs\cos\theta$$

where θ is the angle between the force $\overline{F}$ and the displacement $\overline{s}$ of the mass in question. We see from Figure (a) that the angle between $\overline{F}$ and $\overline{s}$ is 60°:

$$W = Fs\cos 60° = \frac{1}{2}\text{mgs}$$

$$= \frac{1}{2}(5\text{ kg})(9.8\text{ m/sec}^2)(8\text{ m}) = 196\text{ kg-m}^2/\text{sec}^2$$

$$= 196\text{ Joules}$$

Another way to solve this problem is to calculate the difference in gravitational potential energy that the block goes through as it slides 8 m down the incline.

We know that this equals the amount of work that gravity does on the block. As the block slides 8 m down the incline, it falls through a vertical height Δh (see Figure (b)):

$$\Delta h = 8\ (\sin 30°)\text{ m} = 4\text{ m}$$

The gravitational potential energy difference that the block experiences is:

$$W = \Delta E_p = mgh_2 - mgh_1 = mg(h_2 - h_1) = mg\Delta h$$

$$= (5\ kg)(9.8\text{ m/sec}^2)(4\text{ m}) = 196\text{ Joules}$$

5.3   Work Done by a Varying Force

$\overline{F}$ = force acting on a particle

$\overline{r}$ = displacement vector of path of particle

The work is done on a particle as it moves from a point a to a point b is

$$W = \int_a^b \overline{F} \bullet d\overline{r} = \int_a^b F\cos\phi\ dr$$

Work done by a varying force–one dimensional case:

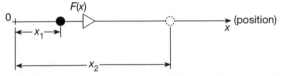

$F(x)$ = The force acting on the particle—a function of position—acts along the x-direction.

Total work done going from x_1 to x_2:

$$W_{12} = \int_{x_1}^{x_2} F(x)\, dx$$

Work done by a spring:

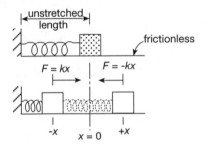

Figure 5.3 Work Done When Stretching a Spring

Force of the spring:

$$\overline{F}(x) = -k\overline{x}$$

k = spring constant

Applied force = $F' = kx$

Work done by the applied force in going from x_1 to x_2:

$$W_{12} = \frac{1}{2}kx_2^2 - \frac{1}{2}kx_1^2$$

If $x_1 = 0$, then

$$W = \frac{1}{2}kx_2^2$$

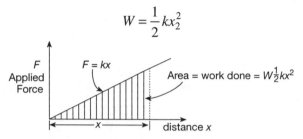

Figure 5.4 Area Under the Force Curve is the Work Done in Stretching a Spring a Distance x

Problem Solving Examples:

Q A small object of weight $\overline{Q}$ hangs from a string of length l, as shown in the figure. A variable horizontal force $\overline{P}$, which starts at zero and gradually increases, is used to pull the object very slowly (so that equilibrium exists at all times) until the string makes an angle θ with the vertical. Calculate the work of the force $\overline{P}$.

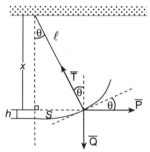

A The object is in equilibrium, meaning that its acceleration is zero and the net force acting on the weight is zero. Consider the forces acting on the object, as shown in the diagram. We can say

$$\Sigma F_x = 0 = P - T\sin\theta \tag{1}$$

$$P = T\sin\theta \tag{2}$$

and
$$\Sigma F_y = 0 = T\cos\theta - Q \tag{3}$$

$$Q = T\cos\theta \tag{4}$$

Dividing equation (2) by equation (4), we get $P = Q\tan\theta$. Since $\overline{P}$ is variable, the work done by it must be found through integration. Recall that work is the integral of the dot product of the force $\overline{F}$ and the displacement vector $d\overline{x}$:

$$W = \int_{x_1}^{x_2} \overline{F} \bullet d\overline{x} = \int_{x_1}^{x_2} F\cos\gamma \, dx$$

where γ is the angle between $\overline{F}$ and $d\overline{x}$. In this case, the force is $\overline{P}$, the differential displacement is $ld\theta$, and the angle between the two is θ. Substituting these expressions, we have

$$W = \int_0^\theta \overline{P} \times ld\overline{\theta} = \int_0^\theta (Q\tan\theta)(\cos\theta)(l) \, d\theta$$

$$= Ql \int_0^\theta \sin\theta \, d\theta = -Ql\cos\theta \Big|_0^\theta = Ql(1-\cos\theta)$$

This result can also be derived using conservation of energy. Since the object's initial and final velocity is zero, kinetic energy is not involved. The change in the object's potential energy must be due completely to the work done on the weight by the force $\overline{P}$. This change in potential energy, ΔPE, is

$$\Delta PE = Qh = Q(l - x)$$

But
$$x = l\cos\theta$$

Therefore, we have

$$\Delta PE = Q(l - l\cos\theta) = Ql(1-\cos\theta)$$

This is equal to the work:
$$W = Ql(1-\cos\theta)$$

Spring-Heel Jack was a legendary English criminal who was never captured because of his ability to jump over high walls and other obstacles which his pursuers were unable to scale. It is believed that he had a powerful spring attached to each shoe for this purpose. Assuming that he weighed 75 kg and that his springs were compressed by 2 cm when he stood on them, by how much did he need to keep his springs compressed on one of his operations in order to be ready to clear a 3 m wall in the event of an emergency?

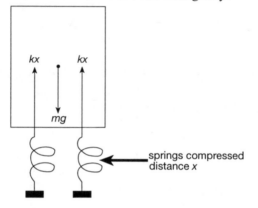

The figure shows an idealized drawing of Spring-Heel Jack. When Jack stands up, the springs are compressed a distance x. When in equilibrium, the net force on Jack is zero. Hence,

$$2kx = mg \quad \text{or} \quad k = \frac{mg}{2x}$$

where m is his mass and k is the spring constant. Thus,

$$k = \frac{75 \text{ kg}}{(2)(0.02 \text{ m})} = \frac{75 \text{ kg}}{0.04 \text{ m}} = 1,875 \text{ kg/m}$$

If Jack wishes to clear a height h while remaining erect, the potential energy stored in the springs must have been sufficient to raise his 75 kg weight through a vertical distance h. But if x was the compression of each spring, then by conservation of energy

$$\frac{1}{2}kx^2 + \frac{1}{2}kx^2 = mgh$$

$$x^2 = \frac{mgh}{k}$$

$$\therefore x^2 = \frac{75\ kg \times 3\ m}{1,875} = 0.12\ m^2$$

$$x = \sqrt{0.12}\ m = 0.35\ m$$

5.4 Power

Power (P) is defined as the rate at which work is performed.

a) Average Power

$$P = \frac{W}{t} \rightarrow \text{units: } \frac{\text{joules}}{\text{sec}} = \text{watts}$$

W = work done during an interval of time t

b) Instantaneous Power

$$P = \frac{dW}{dt} \rightarrow \text{units: } \frac{\text{joules}}{\text{sec}} = \text{watts}$$

or $$P = \overline{F} \times \overline{v}$$

where

$\overline{F}$ = force acting on particle

$\overline{v}$ = velocity vector

If the force is in the direction of the velocity, then

$$P = Fv$$

If the power is constant, then

$$W = Pt$$

Problem Solving Example:

 A constant horizontal force of 10 N is required to drag an object across a rough surface at a constant speed of 5 m/sec. What power is being expended? How much work would be done in 30 min?

Power is the rate of doing work,

$$P = \frac{\Delta W}{\Delta t} = \frac{F\Delta s}{\Delta t}$$

(Note that in this problem the work reduces to the force multiplied by the distance the object is moved.) But

$$\frac{\Delta s}{\Delta t}$$

is just the velocity. Therefore,

$$
\begin{aligned}
P &= Fv \\
&= (10\,\mathrm{N}) \times (5\ \mathrm{m/sec}) \\
&= 50\ \mathrm{J/sec} \\
&= 50\,\mathrm{W}
\end{aligned}
$$

$$
\begin{aligned}
W &= Pt \\
&= (50\,\mathrm{W}) \times \left(\frac{1}{2}\,\mathrm{hr}\right) \\
&= 25\,\mathrm{W\text{-}hr}
\end{aligned}
$$

The work, of course, is done against the force of sliding friction.

5.5   Kinetic Energy

$$K = \frac{1}{2}mv^2 \rightarrow \text{units: } \frac{\text{kg} \times \text{m}^2}{\text{sec}^2} = \text{joules}$$

m = mass of object

v = velocity of object

Problem Solving Example:

   Air consists of a mixture of gas molecules that are constantly moving. Compute the kinetic energy K of a molecule that is moving with a speed of 500 m/s. Assume that the mass of this particle is 4.6×10^{-26} kg.

   The mass of the gas molecule, $m = 4.6 \times 10^{-26}$ kg, and its speed $v = 5 \times 10^2$ m/s, are the known observables. Using the equation:

$$K = \frac{1}{2}mv^2$$

$$K = \left(\frac{1}{2}\right)\left(4.6 \times 10^{-26} \text{ kg}\right)\left(5.0 \times 10^2 \text{ m/s}\right)^2$$

$$= 5.75 \times 10^{-21} \text{ J}$$

5.6   Work–Energy Theorem

The work done by the resultant external force acting on a particle is equal to the change in kinetic energy of the particle.

$$W = K_2 - K_1 = \frac{1}{2}mv_2^2 - \frac{1}{2}mv_1^2 = \Delta K$$

Problem Solving Example:

 Q A single body of mass M in free space is acted on by a constant force $\overline{F}$ in the same direction in which it is moving. Show that the work done by the force is equal to the increase in kinetic energy of the body.

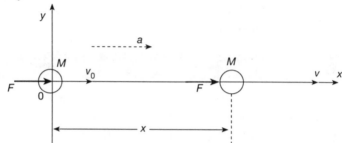

A This is a case of motion in a straight line with constant acceleration. In the figure suppose that at the zero of time the body is at the origin and is moving with a velocity $\overline{v}_0$. Suppose that at time t it has moved through a distance $\overline{x}$ and its velocity has changed to $\overline{v}$. If $\overline{a}$ is the constant acceleration, then we can derive an equation relating v to x.

From the equations:

$$v = v_0 + at \quad \text{and} \quad x = v_0 t + \frac{1}{2} at^2$$

we must eliminate t. Since

$$t = \frac{v - v_0}{a},$$

then,

$$x = v_0 \left(\frac{v - v_0}{a} \right) + \frac{1}{2} a \left(\frac{v - v_0}{a} \right)^2 = \frac{v^2 - v_0^2}{2a}$$

The work done by the force $\overline{F}$ in moving the body from the origin through a distance x is

$$\text{Work done} = Fx = \frac{F\left(v^2 - v_0^2\right)}{2a}$$

But from Newton's second law $\overline{F} = m\,\overline{a}$. So work done

$$= \frac{ma\left(v^2 - v_0^2\right)}{2a}$$

$$= \frac{1}{2}m\left(v^2 - v_0^2\right)$$

$$= \frac{1}{2}mv^2 - \frac{1}{2}Mv_0^2$$

Work done = Final kinetic energy – initial kinetic energy.

Quiz: Work and Energy

1. The diagram below shows a 10 newton force pulling an object up
    a hill at a constant rate of 4 meters per second. How much work is
    done in moving the object from Point *A* to Point *B*?

    (A) 100 J

    (B) 120 J

    (C) 200 J

    (D) 300 J

    (E) 1,000 J

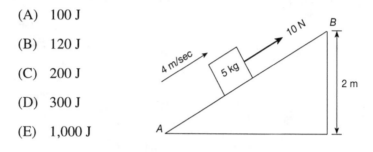

2. A 5 kg cart accelerated at 3 m/s² for 4 seconds. After this time,
    what is its kinetic energy?

    (A) 22.5 J   (D) 360 J

    (B) 60 J   (E) 720 J

    (C) 180 J

3. A box weighing 200 newtons is lifted 2.0 meters by pushing it up
    a ramp with a force of 350 newtons. If 76% of the work applied is
    used to move the box and 24% of the work applied is used to
    overcome the friction, what is the length of the ramp?

    (A) 2.85 meters   (D) 11.4 meters

    (B) 8.7 meters   (E) 15.0 meters

    (C) 9.2 meters

4. A constant net force of 25 N is exerted on a 50 kg cart for 4 seconds. At the end of the 4 seconds, the net force goes to 0 N. If the cart starts from rest, what will be the velocity of the cart after a total of 6 seconds has passed?

    (A) 1.0 m/sec   (D) 2.5 m/sec

    (B) 1.5 m/sec   (E) 3.0 m/sec

    (C) 2.0 m/sec

5. A block resting on an adjustable inclined plane begins to slide when the angle of the plane is 30° with respect to the horizontal. The coefficient of static friction between the block and the plane is

    (A) 0.866.   (D) 0.577.

    (B) 0.700.   (E) 1.00.

    (C) 0.500.

6. Imagine that an object of mass m ($m = 2$ kg) has position vector $\bar{r}$ = $(3t + 5t^3)\,\bar{x}$. Calculate the work done on the particle over the time interval from 0 to 1 s.

    (A) 78 J   (D) 315 J

    (B) 157 J   (E) 393 J

    (C) 235 J

QUESTIONS 7–9 refer to the following diagram.

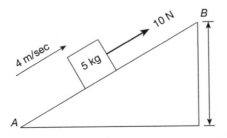

The diagram on the previous page shows a 10 newton force pulling an object up a hill at a constant rate of 4 meters per second.

7. What is the kinetic energy of the object?

(A) 40 J   (D) 0 J

(B) 10 J   (E) 20 J

(C) 100 J

8. How much work is done in moving the object from Point *A* to Point *B*?

(A) 100 J   (D) 300 J

(B) 120 J   (E) 1,000 J

(C) 200 J

9. What is the momentum of the object?

(A) 0 kg m/s   (D) 20 kg m/s

(B) 10 kg m/s   (E) 30 kg m/s

(C) 15 kg m/s

10. A block of weight W is pulled along a horizontal surface at a constant speed V by a force F, which acts at an angle θ with the horizontal, as shown below. The normal force exerted on the block by the surface is equal to

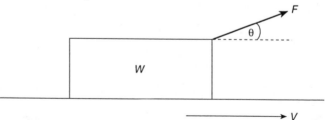

(A) $W - F \sin \theta$.

(D) $W + F \cos \theta$.

(B) $W - F \cos \theta$.

(E) W.

(C) $W + F \sin \theta$.

ANSWER KEY

1.	(A)	6.	(D)
2.	(D)	7.	(A)
3.	(E)	8.	(A)
4.	(C)	9.	(D)
5.	(D)	10.	(A)

CHAPTER 6

Conservation of Energy

6.1 Conservative Forces

A force, or force field, is said to be conservative if any one (and hence, all) of the three following properties are satisfied:

a) The work done by the force on a particle that moves through any round trip is zero.

b) The work done by the force on a particle is independent of the path followed.

c) The kinetic energy of a particle subject to a force returns to its initial value after any round trip.

A force is said to be nonconservative, or dissipative, if any one of the above three conditions are not met.

The force of friction is an example of a nonconservative force.

$\bar{F} = m\bar{g}$

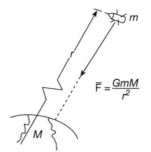

$\bar{F} = \dfrac{GmM}{r^2}$

a) The gravitational force near the earth's surface is an example of a conservative force.

b) The gravitational force that the earth exerts on a satellite is an example of a conservative force.

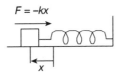

$F = -kx$

x

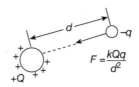

$F = \dfrac{kQq}{d^2}$

c) The elastic force F that the spring exerts on an attached mass is an example of a conservative force.

d) The electrostatic force that is exerted on the $-q$ charge by the $+Q$ charge is an example of a conservative force.

Figure 6.1 Examples of Conservative Forces

Problem Solving Example:

a) Calculate the work done by gravity when a mass of 100 g moves from the origin to $\bar{r} = (50\,\bar{i} + 50\,\bar{j})$ cm. b) What is the change in potential energy in this displacement? c) If a particle of mass M is projected from the origin with speed v_0 at angle θ with the horizontal, how high will it rise?

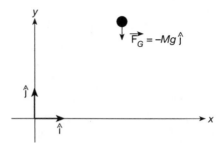

A   a) Let $(\bar{i}, \bar{j})$ be the unit vectors along the horizontal and vertical directions, respectively, as shown in the figure. The gravitational force is

$$\overline{F} \qquad\qquad \overline{F}_G = -Mg\,\bar{j}$$

The work W done by $\overline{F}_G$ is

$$W = \int_{(0,\,0)\text{cm}}^{(50,\,50)\text{cm}} \overline{F}_G \bullet d\bar{r} = -Mg\int_0^{50\text{ cm}} dy \qquad (1)$$

$$W = -Mg\,(50\text{ cm})$$

$$W = -(100\text{ g})\left(980\text{ cm/s}^2\right)(50\text{ cm})$$

$$W = -4.9\times10^6\text{ ergs} \qquad (2)$$

The gravitational force does a negative amount of work. The reason for this is that F_G opposes the upward motion of M from the origin.

b) The definition of potential difference is

$$V_{(50,50)} - V_{(0,0)} = \int_{(0,\,0)}^{(50,\,50)} \overline{F}_G \bullet d\bar{r}$$

From equations (1) and (2),

$$V_{(50,50)} - V_{(0,0)} = 4.9\times10^6\text{ ergs}$$

c) In order to find the maximum height h that the particle attains, we relate the energy at the point of projection ($x = 0, y = 0$) to the energy at $y = h$. This may be done using the principle of energy conservation. Hence,

$$E_f = E_i$$

$$\frac{1}{2} M v_f^2 + V_f = \frac{1}{2} M v_0^2 + V_0$$

We may arbitrarily set $V = 0$ and $y = 0$. Hence, $v_0 = 0$.

$$\frac{1}{2} M v_f^2 + Mgh = \frac{1}{2} M v_0^2 \tag{3}$$

But

$$v_f^2 = v_{x_f}^2 + v_{y_f}^2 \tag{4}$$

$$v_0^2 = v_{x_0}^2 + v_{y_0}^2 \tag{5}$$

because there is no x-component of acceleration. Also, $v_{x_0} = v_{x_f}$ at $y = h$, $v_y = 0$, hence, $v_{x_f} = 0$. Substituting this data in equations (4) and (5),

$$vf^2 = v_{x_0}^2$$

$$v_0^2 = v_{x_0}^2 + v_{y_0}^2$$

Substituting this in equation (3),

$$\frac{1}{2} M v_{x_0}^2 + Mgh^0 = \frac{1}{2} M (v_{x_0}^2 + v_{y_0}^2)$$

or

$$Mgh = \frac{1}{2} M v_{y_0}^2$$

or

$$h = \frac{v_{y0}^2}{2g}$$

But

$$v_{y0} = v_0 \sin \theta,$$

and

$$h = \frac{v_0^2 \sin^2 \theta}{2g}$$

6.2 Potential Energy

A conservative force field has an associated potential energy function U, such that the change in potential energy, ΔU, is equal to the negative of the work done.

$$\Delta U = -W$$

W = work done by the conservative force

Problem Solving Example:

 A 20 kg stone is pushed, on a 30° incline, to the top of a building 30 m tall. By how much does its potential energy increase?

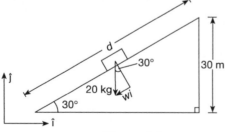

Figure (a)   **Figure (b)**

wi = Component of stone's weight along incline

A The change in a body's gravitational potential energy is the negative of the work done by gravity on the object in displacing it. By definition, this is

$$W = -\int \overline{F}_g \cdot d\overline{r}$$

where F_g, the force of gravity, is

$$\overline{F}_g = -mg\overline{j} \tag{1}$$

The symbol $\overline{j}$ is a unit vector in the positive y-direction (see figure). Now

$$d\overline{r} = dx\overline{i} + dy\overline{j}$$

where $\overline{i}$ is a unit vector in the positive x direction. Then

$$\overline{F}_g \bullet d\overline{r} = -mg\overline{j} \bullet (dx\overline{i} + dy\overline{j}) = -mgdy$$

and

$$W = -\int -mgdy \tag{2}$$

We evaluate (2) over the path of motion of the block. If we take the origin of our coordinate system at the foot of the plane, y varies from 0 to 30 m. Therefore,

$$W = mg\int_0^{30\text{ m}} dy = mg(30\text{ m})$$

$$W = (20\text{ kg})(30\text{ m}) = 600\text{ kg} \times \text{m}$$

6.3   Conservation of Mechanical Energy

The total mechanical energy E of a particle subject to a conservative force remains constant throughout the motion of the particle.

$$K + U = \frac{1}{2}mv^2 + U = E = \text{constant}$$

$$K = \text{kinetic energy} = \frac{1}{2}mv^2$$

$$U = \text{potential energy}$$

$$K_1 + U_1 = K_2 + U_2$$

The kinetic plus potential energy at station 1 is equal to the kinetic plus potential energy at station 2.

These results hold true only if no work is done by nonconservative forces.

Problem Solving Example:

  A delicate machine weighing 175 kg is lowered gently at constant speed down planks 3 m long from the tailboard of a truck 1.5 m above the ground. The relevant coefficient of sliding friction is 0.5. Must the machine be pulled down or held back? If the required force is applied parallel to the planks, what is its magnitude?

The machine is reloaded in the same manner, a force of 1,650 N is being applied. With what velocity does it reach the tailboard? What kinetic energy and what potential energy has it then acquired and how much work has been performed in overcoming friction? What relationship exists between these quantities?

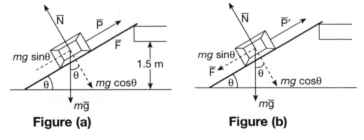

Figure (a)   **Figure (b)**

A The forces acting on the machine are four in number. (1) The weight, $m\,\overline{g}$, acting vertically downward. (2) The normal force exerted by the plane, $\overline{N}$. (3) The frictional force, $\overline{F}$, acting up the plane opposing the motion down it. If μ is the coefficient of sliding friction, this force has magnitude μN. (4) The force $\overline{P}$ is necessary to keep the machine moving with constant speed. In the diagram this is drawn acting up the plane. If the machine has to be pulled down, $\overline{P}$ will be negative.

Resolve the force $m\overline{g}$ into its components along the plane and at right angles to the plane. The forces are in equilibrium since there is no acceleration taking place. Hence, by Newton's second law,

$$N = mg \cos \theta$$

and $$mg \sin \theta = P + \mu N = P + \mu mg \cos \theta$$

$$P = mg ((\sin \theta) - \mu (\cos \theta))$$

Therefore, $P = mg(\sin \theta - \mu\cos \theta)$. But $\sin \theta = 1.5/3.0 = 1/2$ as shown in Figure (a),

$$\therefore \theta = 30°$$

$$\therefore P = 175 \text{ kg} \left(\frac{1}{2} - 0.5 \frac{\sqrt{3}}{2} \right) = 175 \times 0.067 \text{ kg}$$

$$= 11.72 \text{ kg}$$

The machine must be held back with a force of this magnitude.

During the loading process, the forces acting are those shown in Figure (b). Compared with the previous case, (1) and (2) are the same as before; (3) is of the same magnitude but, since it still acts against the motion, its direction is now reversed; and (4) is replaced by the force $\overline{P}$ supplied by the loaders.

There is no tendency to move at right angles to the plane. Thus $N = mg \cos \theta$. The net force up the plane is

$$P' - mg \sin \theta - \mu N = P' - mg (\sin \theta + \mu \cos \theta)$$

$$= 1,650 \text{ N} - 175 \,(9.8) \left(\frac{1}{2} + 0.5 \frac{\sqrt{3}}{2} \right) \text{kg}$$

$$= (1,650 - 1,600)\text{N} = 50 \text{ N}$$

This force is acting on a mass of 175 kg and will produce an acceleration $a = 0.28$ m/s^2 as a result of Newton's second law. The velocity after the machine has traveled 3 m from rest is thus given by the kinematics equations for constant acceleration. In this case, we use the equation

$$v^2 = v^2_0 + 2a(x - x_0)$$

where $x - x_0$ is the distance travelled along the plane, a is the machine's acceleration parallel to the plane, and v_0 is its velocity. Since $v_0 = 0$,

$$v^2 = 2 \times 0.28 \text{ m/s}^2 \times 3 \text{ m}$$

or

$$v^2 = 1.68 \text{ m}^2/\text{s}^2$$

and

$$v = 1.296 \text{ m/s}$$

The kinetic energy at that time is

$$\frac{1}{2} mv^2 = \frac{1}{2} \times 175 \text{ kg} \times 1.68 \text{ m}^2/\text{s}^2 = 247 \text{ N}$$

Or, alternatively, the work-energy theorem tells us the kinetic energy is the net force up the plane times the length of the plane 3m.

$$= 50 \text{ N} \times 3 \text{ m} = 150 \text{ N•m}$$

The potential energy is

$$mgh = 175 \text{ kg} \times 9.8 \text{ m/s}^2 \times 1.5 \text{ m} = 2,572.5 \text{ N•m}$$

The work done in overcoming friction is

$$\overline{\text{F}} \bullet \overline{\text{s}} = \mu N \times 3 \text{ m} = 3 \text{ m} \times \mu mg \cos\theta = 1,212.4 \text{ ft-lb.}$$

The work done by the applied force P′ is

$$\overline{D} \bullet \overline{s} = 330 \text{ lb} \times 8 \text{ ft} = 2,640 \text{ ft-lb}$$

But $27.6 + 1,400 + 1,212.4 = 2,640$. Thus, the work done by the applied force equals the kinetic energy plus the potential energy gained by the machine added to the work done to overcome friction. This is merely a statement of the conservation of energy applied to this problem.

6.4 One-dimensional Conservative Systems

a) The Gravitational System

Force: $\qquad W = mg$

Potential energy: $\qquad U(y) = mgy$

$y = 0$ at surface of earth

Conservation of mechanical energy for gravitation

$$\frac{1}{2}mv_1^2 + mgh_1 = \frac{1}{2}mv_2^2 + mgh_2$$

b) The Spring

Force: $\qquad F = -kx$

Potential energy $\qquad U(x) = \frac{1}{2}kx^2$

$x_0 = 0$ is the position of the end of the string when unstretched.

Conservation of mechanical energy for a spring

$$\frac{1}{2}mv_1^2 + \frac{1}{2}kx_1^2 = \frac{1}{2}mv_2^2 + \frac{1}{2}kx_2^2$$

Problem Solving Example:

Q A block of mass m, initially at rest, is dropped from a height h onto a spring whose force constant is k. Find the maximum distance y that the spring will be compressed. See figure below.

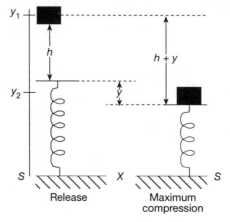

The total fall of the block is $h + y$.

A The general procedure used in solving any problem in mechanics is to calculate all the forces acting on the system and then derive the equation of motion of the system.

An easier way to do mechanics problems involves the use of conservation principles. These laws are not applicable to all problems, but when they are, they simplify the calculation of the solution tremendously.

In this problem, we may use the principle of conservation of energy. We relate the energy of the block before it was released to the block's energy at the point of maximum compression (see figure). At the moment of release, the kinetic energy is zero. At the moment when maximum compression occurs, there is also no kinetic energy.

As shown in the figure, the reference level for gravitational potential energy is the surface S. The initial gravitational potential energy of m is mgy_1. At the point of maximum compression, the potential energy of m is mgy_2. However, at this point, the spring is compressed a distance y and also has elastic potential energy $\frac{1}{2}ky^2$. Hence, equating the energy at the point of release to the energy at the point of maximum compression,

$$mgy_1 = mgy_2 + \frac{1}{2}ky^2$$

$$mg(y_1 - y_2) = \frac{1}{2}ky^2$$

But

$$y_1 - y_2 = h + y$$

and

$$mg(h + y) = \frac{1}{2}ky^2$$

$$y^2 = \frac{2mg}{k}(h + y)$$

$$y^2 - \left(\frac{2mg}{k}\right)y - \frac{2mgh}{k} = 0$$

Therefore, using the quadratic formula to solve for y,

$$y = \frac{1}{2}\left(\frac{2mg}{k} \pm \sqrt{(2mg/k)^2 + (8mgh/k)}\right).$$

6.5 Nonconservative Forces

The work done by friction (a nonconservative force) is equal to the change in internal energy:

$$W_{fr} = \Delta E = -\Delta U_{int}$$

$\Delta E = E_f - E_i$

= final mechanical energy – initial mechanical energy

ΔU_{int} = change in internal (or thermal) energy

Conservation of energy for a system acted upon by conservative and frictional forces

$$\Delta E + \Delta U_{int} = \Delta K + \Delta U + \Delta U_{int} = 0$$

The sum of the mechanical and internal energy of a system acted upon by conservative and frictional (nonconservative) forces remains constant.

Problem Solving Example:

Let us estimate the gravitational energy of the galaxy. We omit from the calculation the gravitational self-energy of the individual stars.

The gravitational energy of an arbitrary system of N particles consists of the sum of the mutual potential energies of each pair of particles. Hence,

$$U = \frac{1}{2} \left\{ \begin{array}{l} (U_{12} + U_{13} + U_{14} + \ldots + U_{1N}) + (U_{21} + U_{22} + U_{23} + \ldots + U_{2N}) \\ + (U_{31} + U_{32} + U_{33} + \ldots + U_{3N}) + \ldots + \\ (U_{N1} + U_{N2} + U_{N3} + \ldots + U_{NN-1}) \end{array} \right\} \quad (1)$$

The terms U_{12}, etc., represent the mutual potential energy of particles 1 and 2. By including sets of terms such as U_{12} and U_{21}, we have double counted, because these represent the mutual potential energies of particles 1 and 2, and particles 2 and 1, respectively. However, these two terms are the same. Hence, the factor $^1/_2$ must be included in (1) to negate the process of double counting. Furthermore, terms such as U_{11} are omitted because they represent the mutual potential energy of particle 1 with particle 1, i.e., they are self energies.

We approximate the gross composition of the galaxy by N stars, each of mass M, and with each pair of stars at a mutual separation of the order of R.

From the definition of potential energy,

$$U_{ij} = \frac{-Gm_im_j}{r_{ij}}$$

where m_i and m_j are the masses of particles i and j, respectively, $G = 6.67 \times 10^{-11}$ N•m²/kg² is their mutual separation. For our case,

$$r_{ij} = R$$

and

$$m_i = m_j = M$$

for all pairs of particles. Then,

$$U_{ij} = \frac{-GM^2}{R} \qquad (2)$$

for any two particles.

Notice that in equation (1), the first parenthesis has $N - 1$ terms of the type of equation 2, the second parenthesis has $N - 1$ terms of this type, and similarly for all parentheses. Altogether, there are $N(N - 1)$ terms of the type of equation 2 in equation 1. Therefore,

$$U = \frac{1}{2}(N)(N-1)\left(\frac{-GM^2}{R}\right)$$

$$U = -\frac{1}{2}\frac{N(N-1)GM^2}{R}$$

If $N \approx 1.6 \times 10^{11}$, $R \approx 10^{21} m$, and $M \approx 2 \times 10^{30}$ kg, then

$$-\frac{1}{2}\left(1.6 \times 10^{11}\right)\left(1.6 \times 10^{11} - 1\right)$$

$$U \approx \frac{\left(6.67 \times 10^{-11} \mathrm{N} \times \mathrm{m}^2/\mathrm{kg}^2\right)\left(2 \times 10^{30}\,\mathrm{kg}\right)^2}{10^{21}\,\mathrm{m}}$$

$$U \approx \frac{-34.15 \times 10^{71}}{10^{21}} J = -34.15 \times 10^{50} J$$

$$U \approx -3.42 \times 10^{51} J$$

6.6   The Conservation of Energy

The total energy—kinetic plus potential plus internal plus all other forms of energy—must remain constant, i.e., energy may be transformed from one kind to another, but it cannot be created or destroyed.

$$\Delta K + \sum \Delta U + \Delta U_{int} + \left(\text{all changes in other forms of energy}\right) = 0$$

Problem Solving Example:

A car coasts down a long hill and then up a smaller one onto a level surface, where it has a speed of 9.8 m/sec. If the car started 67 *m* above the lowest point on the track, how far above this lowest point is the level surface? Ignore friction.

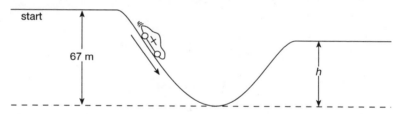

The initial velocity of the car is zero. Since there is no friction, the change in potential energy of the car equals its increase in kinetic energy. On the smaller hill of height *h*, the change in potential energy with respect to the starting point is

$$PE = mg(67 - h)$$

Its kinetic energy is given as

$$KE = \frac{1}{2}mv^2 = \frac{1}{2}m(9.8)^2$$

Equating the two,

$$mg(67-h) = \frac{1}{2}m(9.8)^2$$

$$g(67-h) = 9.8(67-h) = \frac{1}{2}(9.8)^2$$

$$67-h = \frac{1}{2}(9.8)$$

$$h = 67 - 4.9 = 62.1 \text{ m}$$

Therefore, in order for the car to have a speed of 9.8 m/sec, the lower hill must be at a height of 62.1 m above the lowest point on the track.

6.7 The Relationship Between Mass and Energy

a) Redefinition of Mass, m

$$m = \frac{m_0}{\sqrt{1-\left(\frac{v}{c}\right)^2}}$$

m_0 = rest mass

c = speed of light = 3×10^8 m/sec

The mass of a particle is not constant, but instead varies with the v, where v is the velocity relative to the observer.

b) Kinetic Energy

$$K = (m - m_0)c^2$$

c) The Total Energy, E

$$K + \sum U + U_{int} + \ldots + m_0 c^2 = mc^2 = E_{tot}$$

E = total energy = constant

The law of conservation of mass-energy

$$K + \sum \Delta U + U_{int} + \ldots + m_0 c^2 = 0$$

d) Mass-Energy Equivalence Formula

$$E = -\Delta m_0 c^2$$

E = total energy

Δm_0 = change in rest mass

c = speed of light

Problem Solving Examples:

 Find the length of a meter stick ($L = 1$ m) that is moving length-wise at a speed of 2.7×10^8 m/s.

 The length of the stick seems to be smaller when it is moving with respect to the observer. The relativistic formula, giving the contracted length L, is

$$L = L_0 \sqrt{1 - \frac{v^2}{c^2}}$$

where v is the speed of the stick with respect to the observer. Hence,

$$L = (1 \text{ m}) \times \sqrt{1 - \left(\frac{2.7 \times 10^8 \text{ m/s}}{3 \times 10^8 \text{ m/s}}\right)^2}$$

$$L = (1 \text{ m}) \times 0.44 = 0.44 \text{ m}$$

If a 1-kg object could be converted entirely into energy, how long could a 100-W light bulb be illuminated? A 100-W bulb uses 100 J of energy each second.

 Einstein's mass-energy relation can be used to calculate the energy derived from this mass:

$$E = mc^2 = (1 \text{ kg})(3 \times 10^8 \text{ m/s})^2 = 9 \times 10^{16} \text{ J}$$

A 100-watt bulb uses 100 joules of energy in 1 second, by definition of the watt. Hence, for every 100 joules of energy supplied, the bulb remains lit for 1 second. The time, t, the bulb is lighted by 9×10^{16} joules is

$$t = \frac{9 \times 10^{16} \text{ J}}{1 \times 10^2 \text{ J/s}} = 9 \times 10^{14} \text{ s}$$

One year is approximately 3.1×10^7 s, so the time may be written

$$t = \frac{9 \times 10^{14} \text{ s}}{3.1 \times 10^7 \text{ s/year}} = 2.9 \times 10^7 \text{ years}$$

Quiz: Conservation of Energy

QUESTION 1 refers to the following graphs.

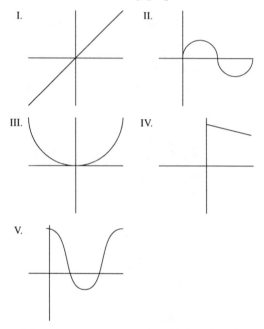

I. II. III. IV. V.

1. Which graph best represents the potential energy in a spring (vertical scale) as a function of displacement from equilibrium (horizontal scale) considering the up and down motion of a small mass on that spring?

(A) I   (D) IV

(B) II   (E) V

(C) III

2. A skier starts from rest at the top of a hill and follows the path shown in the diagram. Assuming no friction, what will be his speed at point *X*?

    (A) $\sqrt{2gh}$ (D) *mgh*

    (B) $\sqrt{2gd}$ (E) *mgd*

    (C) $\sqrt{2g(h-d)}$

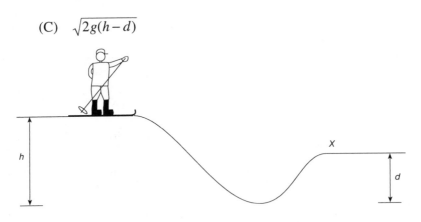

3. As an object is continuously, uniformly accelerated, its velocity levels off, never exceeding the speed of light. Which one of these is true?

    (A) The kinetic energy and momentum also level off.

    (B) The kinetic energy continues to increase but the momentum levels off.

    (C) Both the kinetic energy and the momentum continue to increase.

    (D) The momentum increases but the kinetic energy levels off.

    (E) There is no connection between velocity, kinetic energy, and momentum.

4. What defines a conservative force?

(A) $\oint \overline{\mathbf{F}} \times d\overline{\mathbf{A}} = 0$

(B) The force must be frictional.

(C) The force must be nuclear.

(D) The force must be electromagnetic.

(E) $\oint \overline{\mathbf{F}} \bullet dx = 0$

QUESTIONS 5–8 refer to the following graphs.

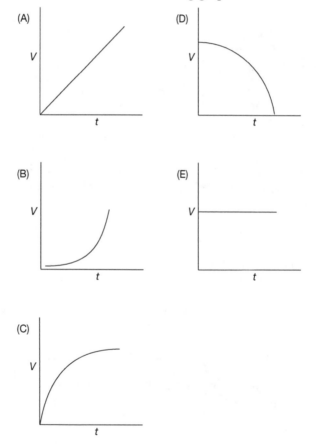

5. A graph that best describes the motion of a mass at rest and pulled by a constant force is

    (A) Graph (A).   (D) Graph (D).

    (B) Graph (B).   (E) Graph (E).

    (C) Graph (C).

6. The graph that best describes the vertical velocity of a mass falling from an airplane through air is

    (A) Graph (A).   (D) Graph (D).

    (B) Graph (B).   (E) Graph (E).

    (C) Graph (C).

7. The graph that best describes the motion of a mass being pulled by a planet with no atmosphere is

    (A) Graph (A).   (D) Graph (D).

    (B) Graph (B).   (E) Graph (E).

    (C) Graph (C).

8. The graph that best describes the motion of a mass pulled by a force that is equal to the frictional force acting on it is

    (A) Graph (A).   (D) Graph (D).

    (B) Graph (B).   (E) Graph (E).

    (C) Graph (C).

9. A force, or force field, is said to be conservative if any one of the following properties are satisfied:

(A) The work done by the force on a particle that moves through any round trip is zero.

(B) The work done by the force on a particle is independent of the path followed.

(C) The kinetic energy of a particle subject to a force returns to its initial value after any round trip.

(D) None of the above.

(E) All of the above.

10. A pendulum of length l is attached to the roof of an elevator near the surface of the earth. The elevator moves upward with acceleration $a = \frac{1}{2}g$. Determine the linear frequency of the pendulum's vibration.

(A) $\dfrac{1}{2}\pi\sqrt{\dfrac{3g}{2l}}$

(B) $\dfrac{1}{2}\pi\sqrt{\dfrac{2g}{3l}}$

(C) $\dfrac{1}{2}\pi\sqrt{\dfrac{g}{l}}$

(D) $\dfrac{1}{2}\pi\sqrt{\dfrac{g}{2l}}$

(E) $\dfrac{1}{2}\pi\sqrt{\dfrac{2g}{l}}$

ANSWER KEY

1.	(C)	6.	(C)
2.	(C)	7.	(B)
3.	(C)	8.	(E)
4.	(E)	9.	(E)
5.	(A)	10.	(C)

CHAPTER 7

The Dynamics of Systems of Particles

7.1 Center of Mass of a System of Particles

a) The x-coordinate

$$x_c = \frac{m_1 x_1 + m_2 x_2 + \ldots + m_n x_n}{m_1 + m_2 + \ldots + m_n} = \frac{\sum\limits_{i=1}^{n} m_i x_i}{\sum\limits_{i=1}^{n} m_i}$$

$$x_c = \frac{\sum\limits_{i=1}^{n} m_i x_i}{M}$$

b) The y-coordinate

$$y_c = \frac{m_1 y_1 + m_2 y_2 + \ldots + m_n y_n}{\sum m_i} = \frac{\sum\limits_{i=1}^{n} m_i y_i}{M} = y_c$$

c) The z-coordinate (for three dimensions)

$$z_c = \frac{\sum\limits_{i=1}^{n} m_i z_i}{M}$$

Here,

$$M = \text{total mass} = \sum\limits_{i=1}^{n} m_i$$

n = number of particles in the system

x_i = x-coordinate of ith particle

y_i = y-coordinate of ith particle

z_i = z-coordinate of ith particle

In vector notation

$$\bar{r}_c = \bar{i} x_c + \bar{j} y_c + \bar{k} z_c$$

Problem Solving Examples:

Q Two masses of 200 gm and 300 gm are separated by a light rod 50 cm in length. The center of mass of the system serves as the origin of a Cartesian coordinate system. The rod lies in the xy plane and makes an angle of 20° with the y axis. Find the inertial coefficients I_{xx} and I_{xy} with respect to the center of mass.

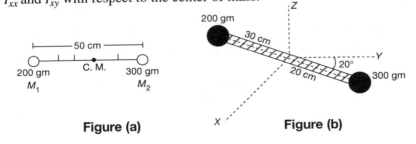

Figure (a)

Figure (b)

Since we must calculate the moment of inertia of the rod about an axis through the center of mass, we must first locate the center of mass. Let us find the distance of the center of mass from the 200-gm mass. By definition of center of mass,

$$R_{c.m.} = \frac{M_1 x_1 + M_2 x_2}{M_1 + M_2}$$

where x_1 and x_2 are the distances of M_1 and M_2 from the origin (in our case, the 200 gm. of mass). Hence, using Figure (a),

$$R_{c.m.} = \frac{(200 \text{ gm})(0) + (300 \text{ gm})(50 \text{ cm})}{500 \text{ gm}}$$

$$R_{c.m.} = 30 \text{ cm}$$

Looking at Figure (b), the Cartesian coordinates of the 200-gm mass (denoted as M_1) are referred to the center of mass as origin,

$$x_1 = (30 \text{ cm}) \sin 20° = (30 \text{ cm})(.342) \approx 10.3 \text{ cm}$$
$$y_1 = (30 \text{ cm}) \cos 20° = (30 \text{ cm})(.940) \approx 28.2 \text{ cm} \qquad (1)$$
$$z_1 = 0$$

The Cartesian coordinates of the 300-gm mass (denoted as M_2) are

$$x_2 = (-20 \text{ cm}) \sin 20° \approx -6.8 \text{ cm}$$
$$y_2 = (-20 \text{ cm}) \cos 20° \approx -18.8 \text{ cm} \qquad (2)$$
$$z_2 = 0$$

Using these values of the coordinates, we proceed to evaluate the inertial coefficients defined by the equations

$$I_{xx} = \sum_i m_i \left(y^2_i + z^2_i \right)$$

$$I_{xx} = -\sum_i m_i x_i y_i$$

From our problem, these reduce to

$$I_{xx} = M_1\left(y^2{}_1 + z^2{}_1\right) + M_2\left(y^2{}_2 + z^2{}_2\right)$$

$$I_{xy} = -M_1 x_1 y_1 - M_2 x_2 y_2$$

From equations (1) and (2)

$$I_{xx} = (200 \text{ gm})\left[(28.2 \text{ cm})^2 + 0\right] + (300 \text{ gm})\left[(-18.8 \text{ cm})^2 + 0\right]$$

$$I_{xx} = 265,080 \text{ gm} \bullet \text{cm}^2 = 2.65 \times 10^5 \text{ gm} \bullet \text{cm}^2$$

$$I_{xy} = -(200 \text{ gm})(10.3 \text{ cm})(28.2 \text{ cm}) - (300 \text{ gm})(-6.8 \text{ cm})(-18.8 \text{ cm})$$

$$I_{xy} = -96,444 \text{ gm} \bullet \text{cm}^2 = -.96 \times \text{gm} \bullet \text{cm}^2$$

Now suppose that the rod rotates about the *x*-axis with angular velocity ω. Find the components of *J*. In general,

$$\begin{pmatrix} J_x \\ J_y \\ J_z \end{pmatrix} = \begin{pmatrix} I_{xx} & I_{xy} & I_{xz} \\ I_{yx} & I_{yy} & I_{yz} \\ I_{zx} & I_{zy} & I_{zz} \end{pmatrix} \begin{pmatrix} \omega_x \\ \omega_y \\ \omega_z \end{pmatrix}$$

where J_x, J_y, and J_z are the vector components of $\bar{J}$, and ω_x, ω_y, and ω_z are the components of $\bar{\omega}$. The quantities I_{xx}, I_{xy}, etc. are the products and moments of inertia of the system we are studying.

In our problem, $\omega_z = \omega_y = 0$ and $I_{zx} = 0$, and

$$\begin{pmatrix} J_x \\ J_y \\ J_z \end{pmatrix} = \begin{pmatrix} I_{xx} & I_{xy} & I_{xz} \\ I_{yx} & I_{yy} & I_{yz} \\ 0 & I_{zy} & I_{zz} \end{pmatrix} \begin{pmatrix} \omega_x \\ 0 \\ 0 \end{pmatrix}$$

Therefore, by the definition of matrix multiplication

$$\begin{pmatrix} J_x \\ J_y \\ J_z \end{pmatrix} = \begin{pmatrix} I_{xx}\,\omega_x & 0 & 0 \\ I_{yx}\,\omega_x & 0 & 0 \\ 0 & 0 & 0 \end{pmatrix}$$

and

$$J_x = I_{xx}\,\omega_x$$
$$J_y = I_{yx}\,\omega_x$$
$$J_z = 0$$

Then

$$\frac{J_y}{J_x} = \frac{I_{yx}}{I_{xx}} = \frac{-.96 \times 10^5}{2.65 \times 10^5} = -.363$$

Q Show that the moment of inertia of a body about any axis is equal to the moment of inertia about a parallel axis through the center of mass plus the product of the mass of the body and the square of the distance between the axes. This is called the parallel-axes theorem.

Prove also that the moment of inertia of a thin plate about an axis at right angles to its plane is equal to the sum of the moments of inertia about two mutually perpendicular axes concurrent with the first and lying in the plane of the thin plate. This is called the perpendicular-axes theorem.

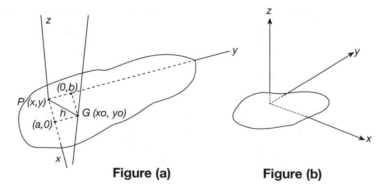

Figure (a)   Figure (b)

A   Let I be the moment of inertia of the body about an arbitrary axis and I_G the moment of inertia about the parallel axis through the center of mass G, the two axes being distance h apart. (See Figure (a) on previous page).

By definition of the center of mass of a body relative to an arbitrary axis through a point P, we obtain

$$I = \sum_i m_i r^2{}_i = \sum_i m_i \left(x_i^2 + y_i^2 \right)$$

$$= \sum_i m_i x_i^2 + \sum_i m_i y_i^2$$

where the sum is carried out over all mass particles, m_i, of the body, and r_i^2 is the distance from P to m_i. Now,

$$x_i = x_i' + a$$

$$y_i = y_i' + b$$

as shown in Figure (a). Here, (x_i', y_i') locates m_i relative to G, the center of mass. Then,

$$I = \sum_i m_i \left(x_i' + a \right)^2 + \sum_i m_i \left(y_i' + b \right)^2$$

$$I = \sum_i m_i \left(x_i'^2 + y_i'^2 \right) + \sum_i m_i \left(a^2 + b^2 \right) + 2a \sum_i m_i x_i' + 2b \sum_i m_i y_i'$$

But $x_i'^2 + y_i'^2 = r_i'^2$ and $a^2 + b^2 = h^2$,

whence $I = \sum_i m_i r_i'^2 + \sum_i m_i h^2 + 2a \sum_i m_i x_i' + 2b \sum_i m_i y_i'$

By definition of the center of mass, however,

$$\sum_i m_i x_i' = \sum_i m_i y_i' = 0,$$

and

$$I = \sum_i m_i r'_i{}^2 + \sum_i m_i h^2 = I_G + Mh^2$$

where

$$M = \sum_i m_i$$

is the net mass of the body. This is the parallel-axes theorem. Although we derived this theorem in two dimensions, it is equally applicable in three dimensions.

Take, in the case of the thin plate, the axes in the plane of the plate as the x- and y-axes, and the axis at right angles to the plane as the z-axis (see Figure (b) on page 122). Then the moment of inertia of the plate about an axis perpendicular to the plate (the z-axis) is

$$I_z = \sum_i m_i r_i^2$$

where r_i locates m_i relative to O. But

$$r_i^2 = x_i^2 + y_i^2$$

where x_i and y_i are the x and y coordinates of m_i. Then

$$I_z = \sum_i m_i x_i^2 + \sum_i m_i y_i^2$$

But

$$\sum_i m_i x_i^2 = I_y \text{ and } \sum_i m_i y_i^2 = I_x,$$

whence $\qquad I_z = I_x + I_y$

This is the perpendicular axes theorem.

7.1.1 Center of Mass in Vector Notation

$$\bar{r}_c = \frac{1}{M} \sum_{i=1}^{n} m_i \bar{r}_i$$

Problem Solving Example:

Q What is the relation between the total energy and the angular momentum for a two-body system, each body executing a circular orbit about the system center of mass?

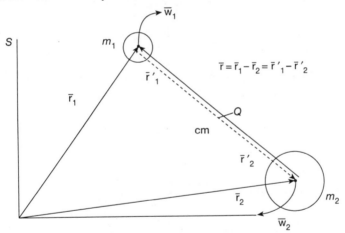

A In order to solve this problem, we must transform the given two-body problem to an equivalent one-body problem. To do this, we find the equation of motion of each mass shown in the figure. Using Newton's second law and his Law of Universal Gravitation, we obtain:

$$\overline{F}_{12} = m_2 \overline{\ddot{r}}_2 = \frac{Gm_1 m_2 (\overline{r}_1 - \overline{r}_2)}{\left| \overline{r}_1 - \overline{r}_2 \right|^3} \tag{1}$$

$$\overline{F}_{21} = m_1 \overline{\ddot{r}}_1 = \frac{Gm_1 m_2 (\overline{r}_1 - \overline{r}_2)}{\left| \overline{r}_1 - \overline{r}_2 \right|^3} \tag{2}$$

where F_{12} is the force exerted on 2 by 1, and similarly for F_{21}. Rewriting equations (1) and (2)

$$\overline{\ddot{r}}_2 = \frac{1}{m_2} \frac{Gm_1 m_2 (\overline{r}_1 - \overline{r}_2)}{\left| \overline{r}_1 - \overline{r}_2 \right|^3} \tag{3}$$

$$\overline{\ddot{r}}_1 = \frac{1}{m_2} \frac{Gm_1 m_2 (\overline{r}_1 - \overline{r}_2)}{\left| \overline{r}_1 - \overline{r}_2 \right|^3} \tag{4}$$

Subtracting equation (3) from (4), and using the figure to realize that $\overline{r}_1 - \overline{r}_2 = \overline{r}$, we obtain

$$\overline{\ddot{r}}_1 - \overline{\ddot{r}}_2 = \overline{\ddot{r}} = -\frac{1}{m_1} \frac{Gm_1 m_2 \overline{r}}{r^3} - \frac{1}{m_2} \frac{Gm_1 m_2 \overline{r}}{r^3}$$

or

$$\overline{\ddot{r}} = -\frac{Gm_1 m_2 \overline{r}}{r^3} \left(\frac{1}{m_1} + \frac{1}{m_2} \right)$$

Defining the reduced mass μ as

$$\frac{1}{\mu} = \frac{1}{m_1} + \frac{1}{m_2} \tag{5}$$

we find

$$\mu \overline{\ddot{r}} = -\frac{Gm_1 m_2 \overline{r}}{r^3} \tag{6}$$

This equation is a one-body equation describing the motion of a particle of mass μ under the influence of a gravitational force.

Now, to further reduce the problem, assume that each mass shown in the figure rotates in a circular orbit with the given angular velocities about Q, the center of mass. Using Newton's second law for each mass,

$$\frac{m_1 v_1'^2}{r_1'} = \frac{G m_1 m_2}{r^2} \tag{7}$$

$$\frac{m_2 v_2'^2}{r_2'} = \frac{G m_1 m_2}{r^2} \tag{8}$$

where the primed variables are measured with respect to the point Q. By definition of the center of mass,

$$m_1 \bar{r}_2' + m_2 \bar{r}_2' = 0$$

or

$$\bar{r}_1' = -\frac{m_2 r_2'}{m_1} \tag{9}$$

Furthermore,

$$\bar{r} = \bar{r}_1' - \bar{r}_2'$$

or

$$\bar{r}_2' = \bar{r}_1' - \bar{r}' \tag{10}$$

Inserting equation (10) in (9) and solving for $\bar{r}_1'$,

$$\bar{r}_1' = -\frac{m_2}{m_1} \bar{r}_2' = -\frac{m_2}{m_1}\left(\bar{r}_1' - \bar{r}\right)$$

$$\bar{r}_1'\left(1 + \frac{m_2}{m_1}\right) = \frac{m_2}{m_1}\bar{r}$$

$$\bar{r}_1' = \frac{m_2/m_1\,\bar{r}}{1 + m_2/m_1} = \frac{m_2\bar{r}}{m_1 + m_2} \tag{11}$$

But using (5),
$$\mu = \frac{m_1 m_2}{m_1 + m_2}$$

Hence, (9) becomes
$$\bar{r}_1' = \frac{\mu}{m_1}\bar{r} \tag{12}$$

Similarly,
$$\bar{r}_2' = \frac{\mu}{m_2}\bar{r} \tag{13}$$

Hence,
$$\bar{r}_1' = \frac{\mu}{m_1}r$$

$$\bar{r}_2' = \frac{\mu}{m_2}r \tag{14}$$

Using (14) and (7) and (8),

$$\frac{m_1^2 v_1'^2}{\mu r} = \frac{Gm_1 m_2}{r^2}$$

$$\frac{m_2^2 v_2'^2}{\mu r} = \frac{Gm_1 m_2}{r^2}$$

or

$$\frac{m_1 v_2'^2}{2} = \frac{\mu Gm_2}{2r}$$

$$\frac{m_2 v_2'^2}{2} = \frac{\mu Gm_1}{2r}$$

Therefore, the net kinetic energy of the system relative to the center of the mass is

(15)

$$T = \frac{1}{2}m_1 v_1'^2 + \frac{1}{2}m_2 v_2'^2 = \frac{\mu G}{2r}(m_1 + m_2) \tag{16}$$

$$T = \frac{Gm_1 m_2}{2r}$$

by definition of μ. The total energy is

$$E = T + V = \frac{Gm_1m_2}{2r} - \frac{Gm_1m_2}{r}$$

$$E = -\frac{Gm_1m_2}{2r} \tag{17}$$

To remove the variable r, we replace it with the angular momentum $\bar{J}$ as follows. The total system angular momentum is (relative to Q).

$$\bar{J} = \bar{r}_1' \times m_1\bar{v}_1' + \bar{r}_2' \times m_2\bar{v}_2'$$

Since $\bar{r}_1'$ and $\bar{v}_1'$ are perpendicular, and similarly for $\bar{r}_2'$ and $\bar{v}_2'$, we obtain

$$J = m_1 r_1' v_1' + m_2 r_2' v_2' \tag{18}$$

From (12) and (13),

$$\bar{v}_1' = \frac{\mu}{m_1}\bar{v} \qquad\qquad \bar{r}_1' = \frac{\mu}{m_1}\bar{r}$$

$$\bar{v}_2' = -\frac{\mu}{m_2}\bar{v} \qquad\qquad \bar{r}_2' = \frac{\mu}{m_2}\bar{r}$$

or

$$\bar{v}_1' = \frac{\mu}{m_1}v \qquad\qquad \bar{r}_1' = \frac{\mu}{m_1}r$$

$$\bar{v}_2' = \frac{\mu}{m_2}v \qquad\qquad \bar{r}_2' = \frac{\mu}{m_2}r \tag{19}$$

Using (19) in (18),

$$J = (m_1)\left(\frac{\mu}{m_1}r\right)\left(\frac{\mu}{m_1}v\right) + (m_2)\left(\frac{\mu}{m_2}r\right)\left(\frac{\mu}{m_2}v\right)$$

$$J = \frac{\mu^2}{m_1} vr + \frac{\mu^2}{m_2} vr = \mu^2 vr \left(\frac{1}{m_1} + \frac{1}{m_2} \right)$$

By definition of μ,

$$J = \mu vr \qquad (20)$$

We now eliminate v in (20) so that we may substitute (20) in place of r in (17). We know from (15) that

$$T = \frac{1}{2} m_1 \dot{v_1}^2 + \frac{1}{2} m_2 \dot{v_2}^2 = \frac{\mu G}{2r} (m_1 + m_2)$$

Substituting for $\dot{v}_1, \dot{v}_2$ from (19) in (15),

$$T = \frac{1}{2} m_1 \left(\frac{\mu^2}{m_1^2} v^2 \right) + \frac{1}{2} m_2 \left(\frac{\mu^2}{m_2^2} v^2 \right) = \frac{\mu G}{2r} (m_1 + m_2)$$

$$T = \frac{\mu^2 v^2}{2m_1} + \frac{\mu^2 v^2}{2m_2} = \frac{\mu G}{2r} (m_1 + m_2)$$

$$T = \frac{\mu^2 v^2}{2} \left(\frac{1}{m_1} + \frac{1}{m_2} \right) = \frac{\mu G}{2r} (m_1 + m_2)$$

By definition of μ,

$$T = \frac{\mu^2 v^2}{2} = \frac{\mu G}{2r} (m_1 + m_2) \qquad (21)$$

Hence,

$$v^2 = \frac{G}{r} (m_1 + m_2) \qquad (22)$$

Using (22) in (20),

$$J = \mu r \sqrt{\frac{G(m_1 + m_2)}{r}}$$

Therefore,
$$r = \frac{J^2}{\mu^2 G(m_1 + m_2)} \qquad (23)$$

Inserting (23) in (17),

$$E = -\frac{Gm_1 m_2}{2}\left(\frac{\mu^2 G(m_1 + m_2)}{J^2}\right)$$

$$E = -\frac{G^2 m_1 m_2 (m_1 + m_2)\mu^2}{2J^2}$$

Finally,

$$m_1 m_2 (m_1 + m_2)\mu^2 = m_1 m_2 (m_1 + m_2)\left(\frac{m_1 m_2}{m_1 + m_2}\right)\mu$$

or $\quad m_1 m_2 (m_1 + m_2)\mu^2 = m_1^2 m_2^2 \mu$

Then
$$E = -\frac{G^2 \mu m_1^2 m_2^2}{2J^2}$$

7.2 Motion of the Center of Mass

Newton's second law for a system of particles:

The center of mass of a system of particles moves as if all the mass of the system were concentrated at the mass center and all the external forces were acting at that point.

$$\overline{F}_{ext} = M\,\overline{a}_c$$

$\overline{a}_c$ = acceleration of the center of mass

$\overline{F}_{ext}$ = the sum of the external forces

M = the total mass of the system

Problem Solving Example:

A rocket, when unloaded, has a mass of 2,000 kg, carries a fuel load of 12,000 kg, and has a constant exhaust velocity of 5,000 km • hr⁻¹. What are the maximum rate of fuel consumption, the shortest time taken to reach the final velocity, and the value of the final velocity? The greatest permissible acceleration is 7 g. The rocket starts from rest at the earth's surface, and air resistance and variations in g are to be neglected.

The problem can be approached using Newton's second law for a system of variable mass.

$$m\frac{d\overline{v}}{dt} = \overline{F}_{ext} + \overline{v}_{rel}\frac{dm}{dt} \tag{1}$$

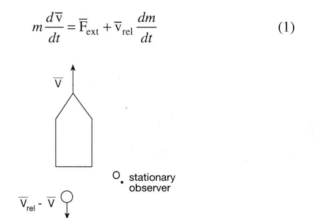

$\overline{F}_{ext}$ is the external force acting on the rocket of mass M. In this case, it is the attractive gravitational force due to the earth (i.e., the weight of the rocket Mg). $\overline{v}$ is the velocity of the rocket relative to a stationary observer O and $\overline{v}_{rel}$ is the relative velocity of the ejected mass with respect to the rocket. The last term is the rate at which momentum is being transferred out of the system by the mass that the system has ejected.

$\overline{F}_{ext} = -mg$, the negative sign indicating that the force acts downward. Also $v_{rel} = -v_r$ where v_r is the velocity of the rocket relative to that of the material ejected, i.e., the reverse of the exhaust velocity. Regarding dm as a very small bit of material ejected in a time dt, and the resulting small increase in velocity of the rocket by dv, then equation (1) becomes

$$m\frac{dv}{dt} = -v_r\frac{dm}{dt} - mg$$

Upon dividing both sides of this equation by m, and multiplying both sides by dt,

$$dv = -v_r\frac{dm}{m} - gdt \qquad (2)$$

Therefore, the acceleration at any time is

$$a = \frac{dv}{dt} = -\frac{v_r}{m}\frac{dm}{dt} - g$$

The velocity v_r is constant and dm/dt must be constant also (for any change in the rate at which mass is ejected would necessarily lead to a change in the velocity of the mass ejected), so that a varies only with m. The smallest value of m gives the greatest value of a; but a cannot exceed 7 g. Therefore,

$$7g = -\left(\frac{5\times10^6\,\text{m-hr}^{-1}}{60\,\text{min-hr}^{-1}\times60\,\text{s-min}^{-1}}\right)\times\frac{1}{2\times10^3\,\text{kg}}\frac{dm}{dt} - g$$

$$\therefore \frac{dm}{dt} = -\frac{8g\times72\times10^5\,\text{kg}}{5\times10^6\,\text{m}\bullet\text{s}^{-1}} = -\frac{8\times9.8\,\text{m}\bullet\text{s}^{-2}\times72\times10^5\,\text{kg}}{5\times10^6\,\text{m}\bullet\text{s}^{-1}}$$

$$= -112.9\,\text{kg}\bullet\text{s}^{-1}$$

where $-dm/dt$ is the maximum rate of fuel consumption. This rate of consumption then equals the total fuel load divided by the time taken to reach the final velocity (T). Thus,

$$T = -\frac{12\times10^3\,\text{kg}}{112.9\,\text{kg}\bullet\text{s}^{-1}} = 106.3\,\text{s} = 1.77\,\text{min}$$

Integrating equation (2), one has

$$v = -v_r \ln m - gt + C.$$

But if m_0 is the total load at time $t = 0$, when $v = 0$, then

$$0 = -v_r \ln m_0 + C$$

$$C = v_r \ln m_0 \text{ and } v = -v_r \ln m - gt + v_r \ln m_0$$

$$\therefore v = v_r \ln \frac{m_0}{m} - gt$$

for

$$\ln \frac{m_0}{m} = \ln m_0 - \ln m$$

The final velocity is thus,

$$v = \frac{5 \times 10^6 \, \text{m} \cdot \text{hr}^{-1}}{60 \times 60 \, \text{s} \cdot \text{hr}^{-1}} \ln \frac{14,000 \, \text{kg}}{2,000 \, \text{kg}} - 9.8 \, \text{m} \cdot \text{s}^{-2} \times 106.3 \, \text{s}$$

$$= (2,703 - 1,042) \text{m} \cdot \text{s}^{-1} = 1,661 \, \text{m} \cdot \text{s}^{-1} = 5,980 \, \text{km} \cdot \text{hr}^{-1}$$

7.3 Work–Energy Theorem for a System of Particles

$$W_c = \Delta K_c = K_{c_f} - K_{c_i} = \frac{1}{2} M V_{c_f}^2 - \frac{1}{2} M V_{c_i}^2$$

W_c = center-of-mass work

K_{c_i}, K_{c_f} = initial and final kinetic energy associated with the center of mass

V_{c_i}, V_{c_f} = initial and final velocity of the center of mass

Problem Solving Example:

In a cloud-chamber photograph, a proton is seen to have undergone an elastic collision, its track being deviated by 60°. The struck particle makes a track at an angle of 30° with the incident proton direction. What mass does this particle possess? (See figure.)

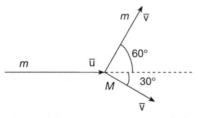

Let the incident proton have mass m and velocity $\bar{u}$, the velocity becoming $\bar{v}$ after scatter. Let the struck particle of mass M acquire velocity $\bar{V}$ after the collision. Then, by the principle of conservation of energy, $\frac{1}{2}mu^2 = \frac{1}{2}mv^2 + \frac{1}{2}MV^2$, and since momentum is conserved both parallel and perpendicular to the original direction of travel of the proton, $mu = mv\cos 60° + MV\cos 30°$ and $mv\sin 60° = MV\sin 30°$. Thus,

$$V = \frac{m}{M}v\frac{\sin 60°}{\sin 30°} = \sqrt{3}\frac{m}{M}v$$

Substituting into the other two equations gives

$$\frac{1}{2}mu^2 = \frac{1}{2}mv^2 + \frac{3}{2}\frac{m^2}{M}v^2$$

and
$$mu = \frac{1}{2}mv + \frac{3}{2}mv$$

$$u^2 = v^2 + 3\frac{m}{M}v^2 = v^2\left(1 + \frac{3m}{M}\right)$$

$$u = \frac{1}{2}v + \frac{3}{2}v = v\left(\frac{1}{2} + \frac{3}{2}\right)$$

$$u^2 = v^2\left(\frac{1}{2} + \frac{3}{2}\right)^2$$

Equating (1) and (2),

$$v^2\left(\frac{1}{2}+\frac{3}{2}\right)^2 = v^2\left(1+\frac{3m}{M}\right)$$

or

$$\frac{1}{4}+\frac{9}{4}+\frac{3}{2}=1+\frac{3m}{M}$$

$$\frac{m}{M}=1$$

and the struck particle must have been a hydrogen nucleus (a proton).

7.4   Linear Momentum of a Particle

$$\overline{p} = m\overline{v} \qquad \text{units:} \quad \frac{\text{kg} \bullet \text{m}}{\text{sec}}$$

$\overline{p}$ = linear momentum of particle

m = mass of particle

$\overline{v}$ = velocity of particle

Problem Solving Example:

A 100-gram marble strikes a 25-gram marble lying on a smooth horizontal surface squarely. In the impact, the speed of the larger marble is reduced from 100 cm/sec to 60 cm/sec. What is the speed of the smaller marble immediately after impact?

The law of conservation of momentum is applicable here, as it is in all collision problems. Therefore, Momentum after impact = Momentum before impact.

Momentum before impact:

$$M_{BI} \times V_{BI} = 100\,\text{gm} \times 100 \text{ cm/sec} = 10,000 \text{ gm-cm/sec}$$
$$(M_{B2} = 25\,\text{gm}, \ V_{B2} = 0)$$

Momentum after impact:

$$= M_{A1} \times V_{A1} + M_{A2} \times V_{A2}$$
$$= 100 \text{ gm} \times 60 \text{ cm/sec} + 25 \text{ gm} \times V_{A2} \text{ cm/sec}$$

Then $10,000 \text{ gm-cm/sec} = 6,000 \text{ gm-cm/sec} + 25 \text{ gm} \times V_{A2}$

whence $\qquad\qquad\qquad V_{A2} = 160 \text{ cm/sec}$

7.4.1 Newton's Second Law

$$\overline{F} = \frac{d\overline{p}}{dt} = \frac{d(m\overline{v})}{dt}$$

where $\overline{F}$ is the net force on the particle.

Problem Solving Example:

Q A sports car weighing 600 kg and traveling at 60 kph fails to stop at an intersection and crashes into a 2,000 kg delivery truck traveling at 45 kph in a direction at right angles to it. The wreckage becomes locked and travels 18.0 m before coming to rest. Find the magnitude and direction of the constant force that has produced this deceleration.

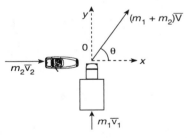

A   Let the sports car be traveling in the positive x-direction and the truck in the positive y-direction. After the collision at the origin, the combined mass travels in a direction inclined at $\theta°$ to the positive x-axis with a velocity $\overline{V}$. Momentum must be conserved in both the x- and y-directions. Therefore, referring to the diagram on the previous page,

$$m_1 v_1 = (m_1 + m_2) V \sin\theta \tag{1}$$

and

$$m_2 v_2 = (m_1 + m_2) V \cos\theta \tag{2}$$

Dividing equation (1) by (2),

$$\frac{(m_1 + m_2) V \sin\theta}{(m_1 + m_2) V \cos\theta} = \frac{m_1 v_1}{m_2 v_2}$$

$$\tan\theta = \frac{m_1 v_1}{m_2 v_2} = \frac{m_1 g v_1}{m_2 g v_2} = \frac{2{,}000 \text{ kg} \times 45 \text{ kph}}{600 \text{ kg} \times 60 \text{ kph}} = 2.5$$

$$\therefore \theta = 68.2°$$

Squaring equations (1) and (2) and then adding them, we get

$$(m_1 + m_2)^2 V^2 \sin^2\theta + (m_1 + m_2)^2 V^2 \cos^2\theta = m_1^2 v_1^2 + m_2^2 v_2^2$$

$$V^2 \left(\sin^2\theta + \cos^2\theta \right) = V^2 = \frac{m_1^2 v_1^2 + m_2^2 v_2^2}{(m_1 + m_2)^2}$$

$$= \frac{m_1^2 g^2 v_1^2 + m_2^2 g^2 v_2^2}{(m_1 g + m_2 g)^2}$$

$$= \frac{(2{,}000 \text{ kg})^2 \times (45 \text{ kph})^2 + (600 \text{ kg})^2 \times (60 \text{ kph})^2}{(2{,}000 + 600 \text{ kg})^2}$$

$$= 1{,}389.9 (\text{kph})^2$$

$$\therefore V = 37.3 \text{ mph} = 54.7 \text{ ft/s}$$

which is the velocity of the combined mass immediately after impact.

The wreckage comes to rest in 18.0 m. Apply the equation for constant acceleration, $v_2 = v_0^2 + 2as$. Here, v_0 is the initial velocity of the locked mass, s is the distance it travels, and a is the applied acceleration. Hence, when $v = 0$, and $s = 18.0$ m and

$$0 = (18.0 \text{ m/s})^2 + 2a \times 18.0 \text{ m}$$

$$\therefore a = -9 \text{ m/s}^2$$

The deceleration due to friction is thus 9 m/s², and since the mass affected is (600 + 2,000), the magnitude of the frictional force is

$$F = 2,600 \text{ kg} \times 9 \text{ m/s}^2 = 23,400 \text{ N}$$

This decelerating force must act in a direction opposite to that in which the wreckage is traveling in order to bring it to rest. Thus, it is a force of 4,443 N. acting at an angle of 68.2° to the negative x-axis. Note that momentum is conserved only during the collision, for, at that time, the collision forces are much greater than the external forces acting (friction), and the latter may be neglected.

7.5 Linear Momentum of a System of Particles

a) Total Linear Momentum

$$\overline{P} = \sum_{i=1}^{n} \overline{p}_i$$

$$= \overline{p}_1 + \overline{p}_2 + \ldots + \overline{p}_n$$

$$= m_1 \overline{v}_1 + m_2 \overline{v}_2 + \ldots + m_n \overline{v}_n$$

$\overline{P}$ = total linear momentum of system

$\overline{p}_i, m_i, \overline{v}_i$ = linear momentum, mass, and velocity of ith particle, respectively

b) Newton's Second Law for a System of Particles (Momentum Form)

$$\bar{F}_{ext} = \frac{d\bar{P}}{dt}; \text{ where } \bar{F}_{ext} = \text{sum of all external forces}$$

Problem Solving Example:

Q   Consider the collision of two particles of mass M_1 and M_2 that stick together after colliding. Let M_2 be at rest initially, and let $\bar{v}_1$ be the velocity of M_1 before the collision. 1) Describe the motion of $M = M_1 + M_2$ after the collision. 2) What is the ratio of the final kinetic energy to the initial kinetic energy? 3) What is the motion of the center of mass of this system before and after collision? 4) Describe the motion before and after the collision in the reference frame in which the center of mass is at rest.

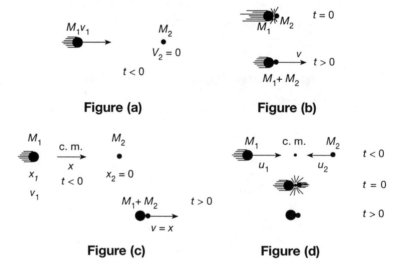

Figure (a): Before collision

Figure (b): During and after collision

Figure (c): Motion of the center of mass

Figure (d): Center of mass system

A 1) The basic principle used in solving a collision problem is the law of conservation of total momentum. This principle may be applied to any collision, so long as there are no external forces (forces due to the outside environment) acting on the system.

From Figures (a) and (b), we see that the initial momentum is $M_1 \bar{v}_1$, and the final momentum is $(M_1 + M_2)\bar{v}$, and we obtain

$$M_1 \bar{v}_1 = (M_1 + M_2)\bar{v}$$

$$\bar{v} = \frac{M_1}{M_1 + M_2} \bar{v}_1$$

Hence, $(M_1 + M_2)$ moves with velocity $\bar{v}$, parallel to $\bar{v}_1$.

2) The kinetic energy K_f after the collision is

$$K_f = \frac{1}{2}(M_1 + M_2)v^2 = \frac{1}{2}(M_1 + M_2)\frac{M_1^2 v_1^2}{(M_1 + M_2)^2}$$

$$K_f = \frac{M_1^2 v_1^2}{2(M_1 + M_2)}$$

The initial kinetic energy k_i is

$$\frac{1}{2}M_1 v_1^2$$

hence,
$$\frac{K_f}{K_i} = \frac{M_1^2 v_1^2}{2(M_1 + M_2)} \times \frac{1}{\frac{1}{2}M_1 v_1^2} = \frac{M_1}{M_1 + M_2}$$

$$\frac{K_f}{K_i} = 1 - \frac{M_2}{M_1 + M_2}$$

The difference $K_i - K_f$ is lost to increased internal motion in the $(M_1 + M_2)$ system (i.e., internal excitations and heat). When a meteorite (M_1) strikes and sticks to the earth (M_2), essentially all the kinetic energy of the meteorite will be lost to heat in the earth. This follows from the fact that if $M_2 \gg M_1$,

$$\frac{K_f}{K_i} = \frac{M_1}{M_1 + M_2} = \frac{1}{1 + \frac{M_2}{M_1}} \approx K_f$$

Hence, $k_f \approx 0$ and all the initial kinetic energy is transformed into heat.

3) The center of mass of a system of particles is a fictitious point whose motion is supposed to describe the trajectory of an imaginary bag which contains all the particles in that system. As the particles move around, the shape and the volume of the bag changes but not its momentum. The interactions of particles among themselves cannot result in a net-resultant force or change in momentum of the bag as a result of the action-reaction principle (for each force exerted on one particle by the others, there is an equal and opposite force exerted by this particle on the others). Furthermore, the collisions of particles with each other conserve the momentum of the colliding particles in each collision and cannot change the total momentum of the bag. In this way, we can view the motion of the center of mass as representing the net effect of only the external forces on the system. If there are no external forces, then the center of mass will not change its velocity, irrespective of the final velocities of the particles.

The position of the center of mass is given by

$$\overline{R}_{cm} = \frac{\text{Sum of all } m_i \overline{r}_1}{M_t}$$

where m_i and $\overline{r}_i$ are the masses and the positions of individual particles, M_t is the total mass. In our problem, let the collision take place at the origin of our coordinate system and at time $t = 0$. After the collision, the center of mass will coincide with the mass $(M_1 + M_2)$ and we have

$$\overline{R}_{cm} = \overline{v}t = \frac{M_1}{M_1 + M_2}\overline{v}_1 t$$

The center of mass velocity is

$$\overline{v}_{cm} = \frac{\text{Sum of all } \overline{p}_i}{M_t}$$

where $\overline{p}_i$ are the individual momenta. For our problem,

$$\overline{v}_{cm} = \overline{v} = \frac{M_1}{M_1 + M_2} \overline{v}_1$$

This expression for the velocity of center of mass will be true for all times, i.e., also before the collision (for which $t < 0$).

4) In this respect, frame $\overline{v}'_{cm} = 0$. (This reference frame is attached to the center of the mass of the system.)

The new velocity of M_1 with respect to this observer will be the velocity of M_1 in the old frame minus the velocity of the c.m. frame with respect to the old frame or

$$\overline{u}_1 = \overline{v}_1 - \overline{v}_{cm} = \frac{M_2}{M_1 + M_2} \overline{v}_1$$

and for M_2

$$\overline{u}_2 = \overline{v}_2 - \overline{v}_{cm} = -\overline{v}_{cm} = -\frac{M_2}{M_1 + M_2} \overline{v}_2$$

This result for $\overline{u}_2$ could be guessed right away. When the observer was in the old frame, M_2 was stationary. As the observer moves with $\overline{v}_{cm}$ with respect to the old frame, then with respect to this observer M_2 will appear to move in the opposite direction with equal speed:

$$\overline{u}_2 = -\overline{v}_{cm}$$

The total momentum $\overline{p}_{total}$ in this system should add up to zero since $\overline{v}_{cm}$ is zero; instead, we have

$$\overline{p}'_{total} = M_1 \overline{u}_1 + M_2 \overline{u}_2$$

$$= \frac{M_1 M_2}{M_1 + M_2} \overline{v}_1 - \frac{M_1 M_2}{M_1 + M_2} \overline{v}_1 = 0$$

The advantage of the center of mass frame is that the total momentum in it is zero.

7.6   Conservation of Linear Momentum

If the sum of the external forces acting on a system is zero, the total linear momentum of the system remains unchanged

$$\overline{P} = \text{constant}$$

For a system of particles:

$$\overline{p}_1 + \overline{p}_2 + \ldots + \overline{p}_n = \overline{P} = \text{constant}$$

or
$$m_1\overline{v}_1 + m_2\overline{v}_2 + \ldots + m_n\overline{v}_n = \text{constant}$$

and the total momentum of the system as a result of collisions is constant.

Problem Solving Example:

 A cart of mass 5 kg moves horizontally across a frictionless table with a speed of 1 m/sec. When a brick of mass 5 kg is dropped on the cart, what is the change in velocity of the cart?

 Assume that the brick has no horizontal velocity when it is dropped on the cart. Its initial horizontal momentum is therefore zero. Since no external horizontal forces act on the system of cart and brick, horizontal momentum must be conserved. We can say, for the horizontal direction,

$$m_c v_{ci} + m_b v_{bi} = m_c v_{cf} + m_b v_{bf}$$

Since the final velocities of the brick and cart are the same,

$$m_c v_{ci} = (m_c + m_b)v_f$$

Substituting values,

$$v_f = \frac{m_c v_{ci}}{m_c + m_b} = \frac{(5\,\text{kg})(1\,\text{m/sec})}{(5\,\text{kg} + 5\,\text{kg})} = .5\,\text{m/sec}$$

The change in velocity of the cart is

$$v_f - v_{ci} = (0.5 - 1.0)\,\text{m/sec} = -0.5\,\text{m/sec}.$$

7.7 Elastic and Inelastic Collisions in One Dimension

When kinetic energy is conserved, the collision is elastic. Otherwise, the collision is said to be inelastic.

a) For an elastic collision,

$$\frac{1}{2}m_1 v_{1\,i}^2 + \frac{1}{2}m_2 v_{2\,i}^2 = \frac{1}{2}m_1 v_{1\,f}^2 + \frac{1}{2}m_2 v_{2\,f}^2$$

b) For an inelastic collision, some kinetic energy is transformed into internal energy. However, linear momentum is still conserved. If the two bodies stick and travel together with a common final velocity after collision, it is said to be completely inelastic. From conservation of momentum, we have

$$m_1 v_{1i} + m_2 v_{2i} = (m_1 + m_2)v_f$$

Problem Solving Example:

 A ball of mass $m_1 = 100$ g traveling with a velocity $v_1 = 50$ cm/sec collides "head on" with a ball of mass $m_2 = 200$ g which is initially at rest. Calculate the final velocities, v_1' and v_2', in the event that the collision is elastic.

In any collision there is conservation of momentum and since this is an elastic collision, kinetic energy is also conserved.

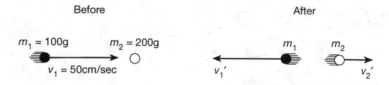

First, we use momentum conservation to write

$$P \text{ (before)} = p \text{ (after)}$$

$$m_1v_1 + m_2v_2 = m_1v_1' + m_2v_2'$$

In order to prevent the equations from becoming too clumsy, we suppress the units (which are CGS throughout); then we have

$$100 \times 50 + 0 = 100v_1' + 200v_2'$$

Dividing through by 100 gives

$$50 = v_1' + 2v_2' \tag{1}$$

From energy conservation, we have (since there is no PE involved and since the collision is elastic)

$$KE \text{ (before)} = KE \text{ (after)}$$

$$\frac{1}{2}m_1v_1{}^2 + \frac{1}{2}m_2v_2{}^2 = \frac{1}{2}m_1v_1'{}^2 + \frac{1}{2}m_2v_2'{}^2$$

$$\frac{1}{2} \times 100 \times (50)^2 + 0 = \frac{1}{2} \times 100 \ v_1'{}^2 + \frac{1}{2} \times 200 \ v_2'{}^2$$

Dividing through by $^{100}/_2 = 50$ gives

$$2,500 = v_1'{}^2 + 2v_2'{}^2 \tag{2}$$

We now have two equations, (1) and (2), each of which contains both of the unknowns, $\bar{v}'_1$ and $\bar{v}'_2$. We can obtain a solution by solving equation (1) for $\bar{v}'_1$.

$$v_1' = 50 - 2v_2' \tag{3}$$

and substituting this expression into equation (2):

$$2,500 = (50 + 2v_2')^2 + 2v_2'^2$$

or

$$2,500 = 2,500 - 200v_2' + 4v_2'^2 + 2v_2'^2$$

From this equation we find

$$6v_2'^2 = 200v_2'$$

so that

$$v_2' = \frac{200}{6} = 33\frac{1}{3} \text{ cm/sec}$$

Substituting this value into equation (3), we find

$$v_1' = 50 - 2 \times 33\frac{1}{3}$$

$$= -66\frac{2}{3} \text{ cm/sec}$$

The negative sign means that after the collision, m_1 moves in the direction opposite to its initial direction (see figure on previous page).

7.8 Collisions in Two and Three Dimensions

Since momentum is linearly conserved, the resultant components must be found and then the conservation laws applied in each direction.

a) The x-component

$$m_1 v_{1i} = m_1 v_{1f} \cos\theta_1 + m_2 v_{2f} \cos\theta_2$$

b) The y-component

$$m_2 v_{2i} = m_1 v_{1f} \sin\theta_1 + m_2 v_{2f} \sin\theta_2$$

where

θ_1 = the angle of deflection, after the collision of mass m_1

θ_2 = the angle of deflection, after the collision of mass m_2

c) For three dimensions, there would be an added z-component and an added angle, θ_3.

For the above cases, i denotes initial value and f denotes final value.

Problem Solving Example:

Q A heavy particle of mass M collides elastically with a light particle of mass m (see the figure below). The light particle is initially at rest. The initial velocity of the heavy particle is $\overline{v}_h = u_h{}^i$; the final velocity is $\overline{w}_h$. If the particular collision is such that the light particle goes off in the forward $(+\overline{i})$ direction, what is its velocity $\overline{w}_1$? What fraction of the energy of the heavy particle is lost in this collision?

A This problem can be solved using the principle of conservation of linear momentum. The linear momentum before collision must equal the linear momentum after collision. The initial momentum p_i of the system is

$$p_i = Mv_h + mv_1 = Mv_h$$

since the smaller mass m is initially at rest. The final momentum p_f of the system is

$$p_f = Mw_h + mw_1$$

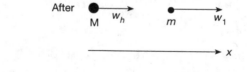

By the conservation of linear momentum, $p_i = p_f$:

$$Mv_h = Mw_h + mw_1$$

Thus,
$$mw_1 = Mw_h - Mv_h$$

$$w_1 = \frac{M}{m}(v_h - w_h)$$

The energy of the heavy particle before the collision is
$$\frac{1}{2}Mv_h^2.$$

After the collision the kinetic energy is
$$\frac{1}{2}Mw_h^2.$$

The fraction of its original kinetic energy that the heavy mass M retains after the collision is

$$\frac{E_f}{E_i} = \frac{\frac{1}{2}Mv_h^2}{\frac{1}{2}Mw_h^2} = \left(\frac{v_h}{w_h}\right)^2$$

where E_i and E_f are the initial and final kinetic energies, respectively.

The fraction of the energy of the heavy particle that is lost in the collision is

$$\text{Fractional Energy Loss} = \frac{E_i - E_f}{E_i} = 1 - \frac{E_f}{E_i} = 1 - \left(\frac{v_h}{w_h}\right)^2$$

Quiz: The Dynamics of Systems of Particles

QUESTION 1 refers to the following before and after diagram. Mass *m* at rest splits into two parts—one with a mass $^2/_3$ *m* and one with a mass $^1/_3$ *m*. After the split, the part with mass $^1/_3$ *m* moves to the right with a velocity *v*.

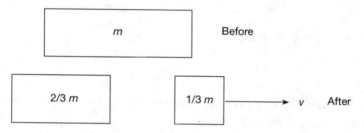

1. The velocity of the $^2/_3$ *m* part after the split is

    (A) $\dfrac{1}{3} v.$ (D) 2*v*.

    (B) $\dfrac{1}{2} v.$ (E) 3*v*.

    (C) $\dfrac{2}{3} v.$

2. A 70 kg person runs with a horizontal velocity of *v* and jumps into a 10 kg raft floating just off shore. Assuming no water resistance, what will be the velocity of the raft and person?

    (A) $\dfrac{1}{7} v$ (B) 7*v*

(C) $\dfrac{8}{7}v$ (D) $\dfrac{1}{8}v$

(E) $\dfrac{7}{8}v$

3. A boy weighing 20 kg riding on a 10 kg cart traveling at 3 m/s jumped off in such a way that he landed on the ground with no horizontal speed. What was the change of speed of the cart?

(A) 1 m/s   (D) 6 m/s

(B) 2 m/s   (E) 9 m/s

(C) 3 m/s

4. A .1 kg ball traveling 20 m/s is caught by a catcher. In bringing the ball to rest, the mitt recoils for .01 second. The absolute value of average force applied to the ball by the glove is

(A) 20 N.   (D) 1,000 N.

(B) 100 N.   (E) 2,000 N.

(C) 200 N.

5. A 30 kg mass traveling due east at 5 m/s collides head on with a 15 kg mass traveling due west at 9 m/s. If the first mass leaves the event due west at 3 m/s, what is the velocity of the second mass?

(A) 0 m/s (at rest)   (D) 13 m/s due east

(B) 5 m/s due west   (E) 25 m/s due east

(C) 7 m/s due east

6. A 2 g bullet initially moving at speed + 200 m/s rips through an apple of mass 100 g and removes a 2 g "chunk." If the speed of the bullet/apple "chunk" system is + 190 m/s, the apple will have a velocity of

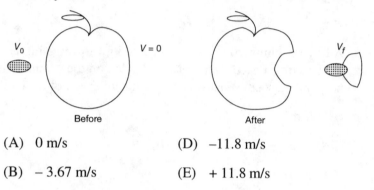

Before   After

(A)   0 m/s   (D)   –11.8 m/s

(B)   – 3.67 m/s   (E)   + 11.8 m/s

(C)   + 3.67 m/s

7. A 10 g bullet is fired into a 2 kg ballistic pendulum as shown in the figure. The bullet remains in the block after the collision, and the system rises to a maximum height of 20 cm. Find the initial speed of the bullet.

(A)   28.0 m/s

(B)   23.8 m/s

(C)   3.98 m/s

(D)   719 m/s

(E)   398 m/s

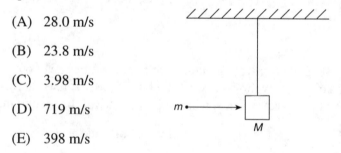

8. A 1 kg mass experiences a net force with the resultant acceleration graphed below. The impulse delivered to the mass is

   (A) 2 N•s.

   (B) 5 N•s.

   (C) 10 N•s.

   (D) $^5/_2$ N•s.

   (E) $^2/_5$ N•s.

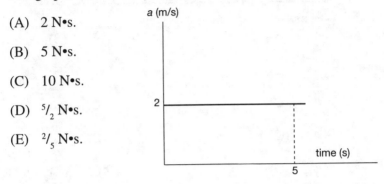

9. A particle in uniform circular motion has

   (A) a greater tangential velocity the closer it is to the axis of rotation.

   (B) an angular acceleration independent of its radial distance.

   (C) an angular velocity independent of its radial distance.

   (D) an angular displacement independent of time.

   (E) an angular velocity dependent upon its radical distance.

10. A car decelerates from 60 m/s to rest in a distance of 240 m. The deceleration is _____ m/s^2.

   (A) 120   (D) 10

   (B) 40   (E) 7.5

   (C) 15

ANSWER KEY

1.	(B)		6.	(B)
2.	(E)		7.	(E)
3.	(D)		8.	(C)
4.	(C)		9.	(C)
5.	(C)		10.	(E)

CHAPTER 8

Rotational Kinematics

8.1 Rotational Motion—The Variables

a) Average Angular Velocity

$$\overline{\omega}_{avg} = \frac{\Delta\theta}{\Delta t} = \frac{\theta_2 - \theta_1}{t_2 - t_1} \rightarrow \text{ units: } \frac{\text{radians}}{\text{second}}$$

$\overline{\omega}_{avg}$ = angular velocity

θ = angular displacement

t = time

b) Instantaneous Angular Velocity

$$\omega = \frac{d\theta}{dt} \rightarrow \text{ units: } \frac{\text{radians}}{\text{second}}$$

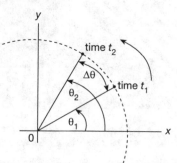

Figure 8.1 Angular Velocity of a Particle

c) Average Angular Acceleration

$$\overline{\alpha}_{avg} = \frac{\omega_2 - \omega_1}{t_2 - t_1} = \frac{\Delta\omega}{\Delta t} = \frac{\text{radians/second}}{\text{second}} = \frac{\text{radians}}{\text{sec}^2}$$

$\overline{\alpha}_{avg}$ = angular acceleration

ω_1, ω_2 = instantaneous angular velocities at times t_1 and t_2

d) Instantaneous Angular Acceleration

$$\alpha = \frac{d\omega}{dt} \rightarrow \text{units: } \frac{\text{radians}}{\text{sec}^2}$$

Problem Solving Examples:

 If a disk is rotating at 1,200 rev/min, what torque is required to stop it in 3.0 min?

 If the disk is to decelerate from 1,200 rev/min to 0 rev/min uniformly, then the angular acceleration (α) will be constant. Hence,

$$\alpha = \text{constant}$$

But

$$\alpha = \frac{d\omega}{dt}$$

where ω is the angular velocity of rotation. Therefore,

$$\alpha = \frac{d\omega}{dt}$$

or

$$d\omega = \alpha dt$$

$$\int_{\omega_0}^{\omega} d\omega = \int_{t=0}^{t=t} \alpha dt$$

$$\omega - \omega_0 = \alpha t \qquad (1)$$

where $\omega = \omega$ at $t = t$ and ω_0 is the initial angular velocity of rotation.

$$\omega_0 = 1,200 \text{ rev/min}$$

Since

$$1 \text{ rev/min} = \frac{1}{60} \text{ rev/sec}$$

$$\omega_0 = 20 \text{ rev/sec}$$

But

$$1 \text{ rev} = 2\pi \text{ radians}$$

and

$$\omega_0 = 40\pi \text{ rad/sec}$$

$$t = 3.0 \text{ min} = 180 \text{ sec}$$

Substituting in equation (1),

$$0 - 40\pi \text{ rad/sec} = \alpha \, (180 \text{ sec})$$

$$\alpha = \frac{-40\pi}{180} \text{ rad/sec}^2$$

This is the acceleration that must be applied to the disk if it is to come to rest in the required time. Because the disk is rotating at a fixed axis, the torque τ is the product of the angular acceleration of the disk and the moment of inertia of the disk about the axis of rotation.

$$\tau = I\alpha$$

$$= \left(38 \, \text{slug-ft}^2\right)\left(\frac{-40\pi}{180}\,\text{rad/sec}^2\right)$$
$$= -26 \, \text{lb-ft.}$$

Hence, a torque of -26 g/m must act on the disk in order to bring it to rest in 3 minutes from a velocity of 1,200 rev/min. The negative sign is consistent with a retarding torque.

 A uniform rod of mass m and length $2a$ stands vertically on a rough horizontal floor and is allowed to fall. Assuming that slipping has not occurred, show that, when the rod makes an angle θ with the vertical

$$\omega^2 = \frac{3g}{2a}(1 - \cos\theta)$$

where ω is the rod's angular velocity.

Also find the normal force exerted by the floor on the rod in this position, and the coefficient of static friction involved if slipping occurs when $\theta = 30°$.

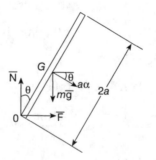

 The forces acting on the rod are the weight $\overline{m}g$ acting downward and the normal force $\overline{N}$ and the frictional force $\overline{F}$ of magnitude μN exerted by the floor at the end 0 in contact with the floor. In order to find ω, we relate the net torque τ on the rod to the rod's angular acceleration α by using

$$\tau = I\alpha$$

Here, I is the rod's moment of inertia. We will then be able to solve for ω.

When one takes moments about 0, the only force producing rotation about 0 is the weight of the rod. Hence,

$$\tau = mga \sin \theta = I_0 \alpha$$

$$I_o = \frac{4}{3} ma^2$$

so

$$d = \frac{3}{4} \frac{g}{a} \sin \theta$$

Here, I_0 is the rod's moment of inertia about 0. Now,

$$\alpha = \frac{d\omega}{dt} = \frac{d\omega}{d\theta} \times \frac{d\theta}{dt} = \omega \frac{d\omega}{d\theta} = \frac{3}{4} \frac{g}{a} \sin \theta$$

$$\omega \frac{d\omega}{d\theta} = \frac{3}{4} \frac{g}{a} \sin \theta$$

$$\int_0^\omega \omega \, d\omega = \int_0^\theta \frac{3}{4} \frac{g}{a} \sin \theta \, d\theta$$

$$\left[\frac{1}{2} \omega^2 \right]_0^\omega = \left[-\frac{3}{4} \frac{g}{a} \cos \theta \right]_0^\theta \quad \text{or} \quad \omega^2 = \frac{3g}{2a}(1 - \cos \theta)$$

The center of gravity G has an angular acceleration about 0, and thus a linear acceleration $a\alpha$ at right angles to the direction of the rod. This linear acceleration can be split into two components: $a\alpha \cos \theta$ horizontally and $a\alpha \sin \theta$ vertically downward. The horizontal acceleration of the center of gravity is due to the force μN and the vertical acceleration is due to the net effect of the forces mg and N. Thus, using Newton's second law, and taking the positive direction downward,

$$mg - N = ma\alpha \sin \theta = \frac{3}{4} mg \sin^2 \theta \qquad (1)$$

and

$$F_{max} = \mu N = ma\alpha \cos \theta = \frac{3}{4} mg \sin \theta \cos \theta \qquad (2)$$

From (1) $\qquad N = mg - \dfrac{3}{4} mg \sin^2 \theta = \dfrac{mg}{4}\left(4 - 3 \sin^2 \theta\right)$

But when $\theta = 30°$, slipping just commences. At this angle F has its limiting, maximum value of F_{max}. We have

$$\mu_s = \frac{F_{max}}{N} = \frac{\dfrac{3}{4} mg \sin \theta \cos \theta}{\dfrac{mg}{4}\left(4 - 3 \sin^2 \theta\right)} = \frac{3 \sin \theta \cos \theta}{\left(4 - 3 \sin^2 \theta\right)}$$

$$= \frac{3 \times \dfrac{1}{2} \times \left(\dfrac{\sqrt{3}}{2}\right)}{4 - \dfrac{3}{4}} = \frac{3\sqrt{3}}{13} = 0.400$$

8.2 Rotational Motion with Constant Angular Acceleration

Similarity Table

Rotational Motion	Linear Motion Equivalent
$\alpha = $ constant	$a = $ constant
$\omega = \omega_0 + \alpha t$	$v = v_0 + a t$
$\theta = \dfrac{\omega_0 + \omega}{2} t$	$x = \dfrac{v_0 + v}{2} t$
$\theta = \omega_0 t + \dfrac{1}{2}\alpha t^2$	$x = v_0 t + \dfrac{1}{2} a t^2$
$\omega^2 = \omega_0^2 + 2\alpha\theta$	$v^2 = v_0^2 + 2 a x$
$\theta_0, \theta = $ initial and final angular displacements	
$\omega_0, \omega = $ initial and final angular velocities	

Problem Solving Example:

Q The angular velocity of a body is 4 rad/sec at time $t = 0$, and its angular acceleration is constant and equal to 2 rad/sec². A line OP in the body is horizontal ($\theta = 0$) at time $t = 0$. a) What angle does this line make with the horizontal at time $t = 3$ sec? b) What is the angular velocity at this time?

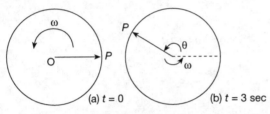

(a) $t = 0$ (b) $t = 3$ sec

A The angular kinematics equations for constant angular acceleration are identical in form to the linear kinematic equations with α corresponding to a, ω to v, and θ to x.

a) Comparable to

$$x = x_0 + \omega_0 t + \frac{1}{2}\alpha t^2,$$

we have $\theta = \theta_0 + \omega_0 t + \frac{1}{2}\alpha t^2$

where θ_0 and ω_0 are the initial angular position and velocity of the body. Since

$$\theta_0 = 0$$

we have $\theta = \omega_0 t + \frac{1}{2}\alpha t^2$

$$= 4\frac{\text{rad}}{\text{sec}} \times 3\,\text{sec} + \frac{1}{2} \times 2\frac{\text{rad}}{\text{sec}^2} \times (3\,\text{sec})^2$$
$$= 21 \text{ radians}$$

$$= 21 \text{ radians } \frac{1 \text{ revolution}}{2\pi \text{ radians}}$$

$$= 3.34 \text{ revolutions}$$

The angle θ is then

$$\theta = 0.34 \times \text{one revolution} = 0.34 \times 360° \approx 122°$$

b)

$$\omega = \omega_0 + \alpha t$$

$$= 4\frac{\text{rad}}{\text{sec}} + 2\frac{\text{rad}}{\text{sec}^2} \times 3\text{sec} = 10\frac{\text{rad}}{\text{sec}}$$

Alternatively,

$$\omega^2 = \omega_0^2 + 2\alpha\theta$$

$$= 4\left(\frac{\text{rad}}{\text{sec}}\right)^2 + 2 \times 2\frac{\text{rad}}{\text{sec}^2} \times 21 \text{ rad}$$

$$= 100\frac{\text{rad}^2}{\text{sec}^2},$$

$$\omega = 10\frac{\text{rad}}{\text{sec}}$$

8.3 Relation Between Linear and Angular Kinematics for a Particle in Circular Motion

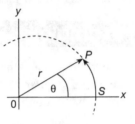 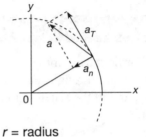

P = particle r = radius

S = arc length a_T = tangential acceleration

q = angle a_n = normal acceleration

Figure 8.2

a) Velocity – Angular Velocity

$$v = \omega r$$

v = linear velocity

ω = angular velocity

r = radius of path of the particle

b) Acceleration – Angular Acceleration

$$a_T = \alpha r$$

a_T = tangential component of acceleration

$$a_n = \frac{v^2}{r} = \omega^2 r$$

a_n = normal component of acceleration

Note that

$$\omega^2 r = \frac{v^2}{r}$$

equals centripetal acceleration in uniform circular motion.

Problem Solving Example:

Q Ignoring the motion of the earth around the sun and the motion of the sun through space, calculate a) the angular velocity, b) the velocity, and c) the acceleration of a body resting on the ground at the equator.

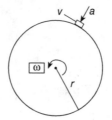

View From the North Pole

A Because of the rotation of the earth, the body at the equator moves in a circle whose radius is equal to the radius of the earth (see figure).

$$r = \text{radius of earth}$$

$$= 6.37 \times 10^6 \text{ meters}$$

We are going to use the MKS system of units. One revolution, which is 2π radians, takes one day or

$$\frac{24 \text{ hr}}{1 \text{ day}} \times \frac{60 \text{ min}}{1 \text{ hr}} \times \frac{60 \text{ sec}}{1 \text{ min}} = 24 \times 60 \times 60 \frac{\text{secs}}{\text{day}}$$

a)   Since the frequency of revolution is $f = {}^1/_T$, where T is the period (the time for one revolution), then

$$2\pi f = \omega = \frac{2\pi}{T}$$

This equals the number of radians traveled per unit time, or the angular velocity ω.

$$\omega = \frac{2\pi}{24 \times 60 \times 60}$$

$$= 7.27 \times 10^{-5} \text{ radians per second}$$

b) The linear velocity is, by definition

$$v = \omega r$$

$$= \left(7.27 \times 10^{-5}\right) \times \left(6.37 \times 10^{6}\right) \frac{\text{rad}}{\text{s}} \bullet \text{m}$$

$$= 4.64 \times 10^{2} \text{ m/sec}$$

Since 1 mph = 0.447 m/sec

$$v = \left(4.64 \times 10^{2} \text{ m/sec}\right)\left(\frac{1}{.447} \frac{\text{sec}}{\text{m}}\right)$$

$$= 1,040 \text{ mph}$$

c) The acceleration toward the center of the earth is, since the motion is circular,

$$a = \frac{v^{2}}{r}$$

$$= \frac{\left(4.64 \times 10^{2} \text{ m/sec}\right)^{2}}{\left(6.38 \times 10^{6} \text{ m}\right)}$$

$$= 3.37 \times 10^{-2} \text{ m/sec}^{2}$$

Quiz: Rotational Kinematics

1. A rotating object which suddenly contracts to a smaller radius rotates with a higher angular velocity because

    (A) smaller objects turn more quickly than larger ones.

    (B) the object's density must increase.

    (C) the rotational inertia of smaller objects is greater.

    (D) the angular velocity of rotating objects must remain the same.

    (E) angular momentum must be conserved.

2. The wheel of an automobile spinning at 180 rev/min begins to experience a 10 rad/sec^2 angular acceleration. What is the angular velocity of the wheel after 5 seconds?

    (A) 182 rev/min   (D) 477 rev/min

    (B) 230 rev/min   (E) 657 rev/min

    (C) 275 rev/min

3. Calculate the centripetal force required to keep a 4 kg mass moving in a horizontal circle of radius 0.8 m at a speed of 6 m/s. ($\bar{r}$ is the radial vector with respect to the center.)

    (A) 39.2 tangent to the circle

    (B) −30.0 N tangent to the circle

    (C) 144.0 N $\bar{r}$

    (D) −180 N $\bar{r}$

    (E) 180 N $\bar{r}$

4. If an object is moving in circular motion due to centripetal force, F, and the radius of its circular motion is then doubled, the new force then becomes

(A) $2F$.

(D) F^2.

(B) F.

(E) $\dfrac{1}{F}$.

(C) $\dfrac{F}{2}$.

5. Rotational motion always results in the absence of an unbalanced

(A) force.

(B) torque.

(C) cross-product.

(D) tension.

(E) None of these.

6. Centripetal force accelerates you

(A) toward the center because your velocity is changing in magnitude.

(B) toward the center because your velocity is changing in direction.

(C) tangentially because your velocity is changing in magnitude.

(D) away from the center because your velocity is changing in magnitude.

(E) away from the center because your velocity is changing direction.

7. The critical velocity needed to complete a vertical circle can be computed using

(A) $\sqrt{rg}$.

(B) $\sqrt{2rg}$.

(C) $\dfrac{\sqrt{rg}}{2}$.

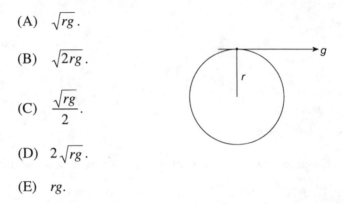

(D) $2\sqrt{rg}$.

(E) rg.

8. A permanent magnet alloy of samarium and cobalt has a magnetization $M = 7.50 \times 10^5$ J/T•m^3. Consider two magnetized spheres of this alloy each 1 cm in radius and magnetically stuck together with unlike poles touching. What force must be applied to separate them?

(A) 74 N   (D) 111 N

(B) 18.5 N   (E) 9.3 N

(C) 37 N

9. A wheel 4 m in diameter rotates with a constant angular acceleration $\alpha = 4$ rad/s^2. The wheel starts from rest at $t = 0$s where the radius vector to point P on the rim makes an angle of 45° with the x-axis. Find the angular position of point P at arbitrary time t.

(A) 45 degrees

(B) $45 + 2t^2$ degrees

(C) $45 + 114.6t^2$ degrees

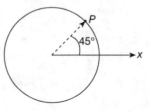

(D) $229.2t^2$ degrees

(E) $4t^2$ degrees

10. A wheel 4 m in diameter rotates on a fixed, frictionless horizontal axis, about which its moment of inertia is 10 kg•m². A constant tension of 40 N is maintained on a rope wrapped around the rim of the wheel. If the wheel starts from rest at $t = 0$s, find the length of rope unwound at $t = 3$s.

(A) 36.0 m

(B) 72.0 m

(C) 18.0 m

(D) 720 m

(E) 180 m

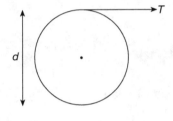

ANSWER KEY

1.	(E)	6.	(B)
2.	(E)	7.	(A)
3.	(D)	8.	(C)
4.	(C)	9.	(C)
5.	(D)	10.	(B)

CHAPTER 9

Rotational Dynamics

9.1 Torque

a) Vector Equation

$$\tau = \bar{r} \times \overline{F}$$

This indicates that the torque is formed by the component of force perpendicular to $\bar{r}$.

b) Scalar Equation

$$\tau = rF \sin \theta$$

τ = torque

r = radius

F = applied force

θ = angle formed by $\bar{r}$ and $\overline{F}$

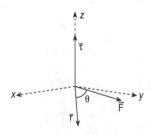

Figure 9.1 Torque

Problem Solving Example:

Q A force $F = 10$ N in the +y-direction is applied to a wrench which extends in the +x-direction and grasps a bolt. What is the resulting torque about the bolt if the point of application of the force is 30 cm (0.3) m away from the bolt?

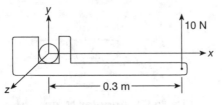

A Torque is calculated from the relation

$$\bar{\tau} = \bar{r} \times \bar{F}$$

where $\bar{\tau}$ stands for torque, $\bar{F}$ stands for the force, and $\bar{r}$ denotes the distance from the origin about which the torque is calculated, of the point of application of the force. In this problem we use the bolt as our origin about which we calculate the torque (see the figure above). Then

$$\bar{\tau} = 0.3 \text{ m } \bar{r} \times 10 \text{ N}\bar{j} = 3 \text{ N} \cdot \text{m} \quad (\bar{i} \times \bar{j}) = 3 \text{ N} \cdot \text{m } \bar{k}$$

where $\bar{i}$, $\bar{j}$, and $\bar{k}$ are the unit vectors in the +x, +y, and +z directions, respectively.

9.2 Angular Momentum

a) Vector Equation

$$\bar{\ell} = \bar{r} \times \bar{p}$$

b) Scalar Equation

$$\ell = rp \sin \theta$$

ℓ = angular momentum

r = radius

p = linear momentum

θ = angle formed by $\bar{r}$ and $\bar{p}$

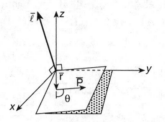

Figure 9.2 Angular Momentum

c) Relation Between Torque and Angular Momentum

$$\bar{\tau} = \frac{d\bar{\ell}}{dt}$$

This vector equation is equivalent to three scalar equations:

$$\tau_x = \frac{d\ell_x}{dt}, \qquad \tau_y = \frac{d\ell_y}{dt}, \qquad \tau_z = \frac{d\ell_z}{dt}$$

x-component y-component z-component

Problem Solving Example:

 A satellite of mass m moves around the earth as shown (actually, the path is an ellipse). Which instantaneous velocity is greater, v_p (at point P) or v_A (at point A)?

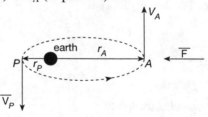

A Consider the earth as a fixed object and neglect the influence of the sun and other planets.

The angular momentum of the satellite around the earth, L, is given by

$$\overline{L} = \overline{r} \times m\overline{v}$$

where $\overline{r}$ is the vector from the earth to the satellite, and $\overline{v}$ is the velocity of the satellite. Since $\overline{v}$ and $\overline{r}$ are perpendicular

$$L = mvr$$

However, $$T = \frac{dL}{dt} \tag{1}$$

where the torque T is defined as

$$\overline{T} = \overline{r} \times \overline{F}$$

$\overline{F}$ is the gravitational force on the satellite keeping it in its orbit. (It is due to the mass of the earth.) Since the angle between $\overline{F}$ and the radius of vector $\overline{r}$ is $0°$, we have

$$T = Fr \sin 0° = 0$$

Therefore, by equation (1), L of the satellite is constant in time. At time t_1 the particle is at A and at time t_2 it is at P. Hence, the angular momentum at the two points must be the same. Or

$$L = mv_A r_A = mv_P r_P$$

Since $r_p < r_A$ we must then have $v_p > v_A$.

The velocity is greatest when the satellite is nearest the earth; this point is called the pedigree (labelled P in the diagram). The velocity is least at the farthest point from the earth—the apogee (A) of the orbit.

9.3 Kinetic Energy of Rotation and Rotational Inertia

a) Kinetic Energy

$$K = \frac{1}{2}mr^2\omega^2$$

since $r^2\omega^2 = v^2$

K = kinetic energy

m = mass

r = radius

ω = angular velocity

b) Total Kinetic Energy

$$K = \frac{1}{2}\left(m_1r_1^2 + m_2r_2^2 + \ldots + m_nr_n^2\right)\omega^2$$

$$K = \frac{1}{2}\left(\Sigma m_nr_n^2\right)\omega^2$$

c) Moment of Inertia

$$I = \Sigma m_nr_n^2$$

The equation for kinetic energy becomes.

$$K = \frac{1}{2}I\omega^2$$

Problem Solving Example:

Q A uniform cylinder rolls from rest down the side of a trough whose vertical dimension y is given by the equation $y = Kx^2$. The cylinder does not slip from A to B, but the surface of the trough is frictionless from B on toward C. (See figure on next page). How far will the cylinder ascend toward C? Under the same conditions, will a uniform sphere of the same radius go farther or less far toward C than the cylinder?

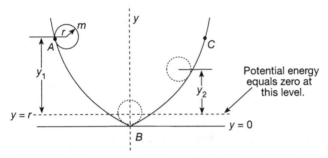

A   Since we do not know the actual frictional force acting on the cylinder, we cannot use dynamical methods to solve for the final position of the cylinder. Our only other recourse is to use the principle of conservation of energy to relate the energy of the cylinder at point A to its energy at point C. By doing this, we will find an expression for the final position of the cylinder, and the problem will be solved.

Since friction acts along path AB, but not along path BC, we apply the principle of energy conservation in two steps. First, we relate the cylinder's energy at points A and B. Then, using the data obtained from the first step, we relate the cylinder's energy at points B and C.

The cylinder begins from rest at point A, and therefore has only potential energy. Taking the reference level for potential energy at $y = r$ (see figure), we obtain

$$E_A = mgy_1 \qquad (1)$$

for the cylinder's energy at A.

In traveling from A to B, friction is present. However, this force does no work because we are given the fact that the cylinder is not slipping. By definition, this means that the velocity of the contact point of cylinder and surface is zero. Under these conditions the velocity of the cylinder's center of mass, v, is related to the angular velocity by

$$V = \omega r \qquad (2)$$

where r is the cylinder radius. Hence, the energy of the cylinder at B is

$$E_B = \frac{1}{2}mv^2 + \frac{1}{2}I\omega^2 \tag{3}$$

where I is the cylinder's moment of inertia. The first term represents the cylinder's translational motion, and the last term represents its rotational motion.

Now, in going from B to C, no friction acts. As a result, the cylinder cannot roll, and the rotational motion it had at B is preserved throughout the trip to C. At C, the cylinder's center of mass has no velocity, but it is still spinning, with angular velocity, ω, about its center of mass. The energy at C is then

$$E_C = mgy_2 + \frac{1}{2}I\omega^2 \tag{4}$$

By the principle of conservation of energy

$$E_A = E_B$$

and $$E_B = E_C.$$

Therefore, using equations (1), (3), and (4)

$$mgy_1 = \frac{1}{2}mv^2 + \frac{1}{2}I\omega^2 \tag{5}$$

$$mgy_2 + \frac{1}{2}I\omega^2 = \frac{1}{2}mv^2 + \frac{1}{2}I\omega^2 \tag{6}$$

From (2) $$\omega = \frac{v}{r}$$

Substituting this in (5)

$$mgy_1 = \frac{1}{2}mv^2 + \frac{1}{2}I\frac{v^2}{r^2} \tag{7}$$

Solving for v^2

$$mgy_1 = \frac{v^2}{2}\left(m + \frac{I}{r^2}\right)$$

or

$$v^2 = \frac{2mgy_1}{\left(m + \frac{I}{r^2}\right)} \tag{8}$$

From (6)

$$mgy_2 = \frac{1}{2}mv^2 \tag{9}$$

We may eliminate v^2 from (9) by substituting (8) in (9), whence

$$mgy_2 = \frac{1}{2}m\left\{\frac{2mgy_1}{\left[\left(m + \frac{I}{r^2}\right)\right]}\right\}$$

then

$$y_2 = \frac{my_1}{\left(m + \frac{I}{r^2}\right)}$$

For a cylinder,

$$I = \frac{1}{2}mr^2$$

and

$$y_2 = \frac{my_1}{m + \frac{1}{2}mr^2/r^2} = \frac{my_1}{\frac{3}{2}m}$$

$$y_2 = \frac{2}{3}y_1$$

For a sphere, the above analysis still holds. Since

$$I = \frac{2}{5}mr^2,$$

$$y_2 = \frac{my_1}{m + \frac{2}{5}r^2/r^2} = \frac{my_1}{\frac{7}{5}m}$$

$$y_2 = \frac{5}{7}y_1$$

Hence, the sphere travels further.

9.4 Some Rotational Inertias

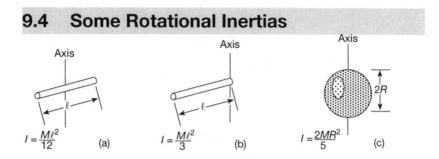

Figure 9.3 The rotational inertia for (a) a thin rod about an axis through its center, perpendicular to the length, (b) a thin rod about an axis through one end, perpendicular to the length, and (c) a solid sphere about any diameter.

Problem Solving Example:

 A thin, rigid rod of weight W is supported horizontally by two props as shown in Figure (a). Find the force F on the remaining support immediately after one of the supports is kicked out.

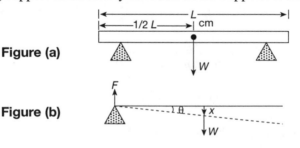

Figure (a)

Figure (b)

 The moment the support is kicked out, the rod starts rotating about the other support as the free end of the support falls (Figure (b)).

Let x be the displacement of the center of mass of the rod. Immediately after the kicking of the support, x is very small and is vertical. In this case

$$\frac{d^2x}{dt^2}$$

becomes the downward acceleration of the center of mass:

$$m\frac{d^2x}{dt^2} = W - F \tag{1}$$

The torque on the rod about the remaining support is

$$\tau = W\frac{L}{2} = I\frac{d^2\theta}{dt^2}$$

where I is the moment of inertia with respect to the axis of rotation. The moment of inertia of a thin rod with respect to an end is known to be $1/3\,mL^2$; hence,

$$\frac{1}{2}WL = \frac{1}{3}mL^2\frac{d^2\theta}{dt^2}$$

or

$$\frac{d^2\theta}{dt^2} = \frac{3}{2}\frac{W}{mL}$$

For small x,

$$x \approx \frac{L}{2}\theta$$

or

$$\frac{d^2x}{dt^2} \sim \frac{L}{2}\frac{d^2\theta}{dt^2}$$

From (1) and (2),

$$m\frac{L}{2}\frac{d^2\theta}{dt^2} = m\frac{L}{2}\frac{3W}{2mL} = W - F$$

$$\frac{3}{4}W = W - F$$

giving

$$F = W - \frac{3}{4}W = \frac{1}{4}W$$

9.5   Rotational Dynamics of a Rigid Body

a)

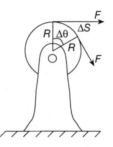

Figure 9.4 Rotational Dynamics of a Rigid Body

$$\Delta w = \tau \Delta \theta \qquad \text{units: joules}$$

w = work done

$\tau = FR$ is the torque, τ, due to the force F

b) If torque is constant while angle changes by a finite amount from θ_2 to θ_1,

$$w = \tau(\theta_2 - \theta_1)$$

c) Power, P

$$P = \tau \omega$$

ω = angular velocity

d) Angular Momentum, L

$$L = I\omega$$

e) Torque, τ

$$\tau = I\alpha$$

Problem Solving Example:

Q A missile is fired radially from the surface of the earth (of radius 3.4×10^6 m) at a satellite orbiting the earth. The satellite appears stationary at the point where the missile is launched. Its distance from the center of the earth is 25.4×10^6 m. Will the missile actually hit the satellite?

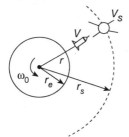 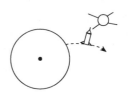

Figure (a) View of Outside Observer   **Figure (b) View from Earth**

A To an observer outside the planet (see Figure (a)), the earth and the missile (of mass m) on the surface are rotating about the axis of the planet with angular velocity ω_0. When the missile is fired radially from the surface, its distance from the center of the earth increases and thus its moment of inertia ($I = mr^2$) about the rotation axis increases also. There are no forces with a moment about the rotation axis acting on the missile (the gravitational force of attraction acting on it is exerted along the rotational axis and has no moment). The net torque Γ acting on the missile is then zero. According to the rigid body analogue of Newton's Second Law, if L is the magnitude of the angular momentum of the missile, then

$$\Gamma = \frac{dL}{dt}$$

but $\Gamma = 0$ and

$$0 = \frac{dL}{dt}$$

or L = constant at all times. But $L = mvr$, where v is the tangential velocity of the missile. Since $v = \omega r$, where ω is the angular velocity of the missile, then $L = m\omega r^2 = I\omega$. Thus, since L and m are constant at all times, as the missile moves farther away from the earth and closer to the satellite (i.e., r increases), then ω must decrease.

We are given the satellite appears stationary at the point where the missile is fired. Thus, the radius vector passing through the launching pad and the satellite continues to rotate with angular velocity ω_0. The missile has an angular velocity ω which drops more and more from the value ω_0 as the missile rises. At the height of the satellite, the moment of inertia of the rocket about the axis of rotation is

$$I_1 = mr^2{}_s = m \left(25.4 \times 10^6 \, \text{m}\right)^2$$

whereas, at the launching pad, its moment of inertia is only

$$I_2 = mr^2{}_e = m \left(3.4 \times 10^6 \, \text{m}\right)^2$$

Thus, finally, if ω is the satellite's angular velocity at height r_s and ω_0 its angular velocity at launching (equal to that of the earth), then

$$I_1\omega = L = I_2\omega_0$$

$$\frac{\omega}{\omega_0} = \frac{I_2}{I_1} = \frac{mr^2{}_e}{mr^2{}_s} = \frac{\left(3.4 \times 10^6 \, \text{m}\right)^2}{\left(25.4 \times 10^6 \, \text{m}\right)^2} = 0.018$$

The missile thus moves farther and farther from the vertical as it rises and will miss the satellite (unless the missile is fitted with a homing device).

To an observer on the planet, the departure of the missile from the vertical is, of course, also observed and is explained in terms of the Coriolis force associated with a rotating frame of reference.

9.6 Rolling Bodies

a) Rotational Inertia

Rolling Cylinder or Disk

$$I_P = I_{cm} + MR^2$$

I_p = rotational inertia about axis through P

I_{cm} = rotational inertia about axis through C

M = mass

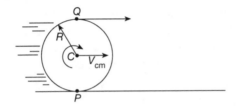

R = radius

Note: The point Q on the cylinder moves with twice the linear velocity of C because it is twice as far from P.

Figure 9.5 Rolling cylinder with mass, M, and velocity of center of mass, V_{cm}

b) Kinetic Energy

$$K = \frac{1}{2} I_p \omega^2$$

or

$$K = \frac{1}{2} I_{cm} \omega^2 + \frac{1}{2} MR^2 \omega^2$$

or

$$K = \frac{1}{2} I_{cm} \omega^2 + \frac{1}{2} MV^2_{cm}$$

Problem Solving Example:

Q A solid cylinder 30 cm in diameter at the top of an incline 2.0 m high is released and rolls down the incline without loss of energy due to friction. Find the linear and angular speeds at the bottom.

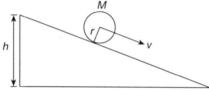

A This problem can be solved using the conservation of energy principle. The cylinder initially at rest at the top of the incline has only gravitational (potential) energy. Taking the bottom of the incline as the zero level of the potential energy (see the figure), we get

$$E_p = mgh$$

where m is the mass of the cylinder, g is the acceleration of gravity, and $h = 2.0$ m is its height above ground level. When the cylinder reaches the bottom of the incline, all of its energy will be kinetic:

$$E_k = \frac{1}{2}mv^2 + \frac{1}{2}I\omega^2$$

where v is the cylinder's linear velocity, I its moment of inertia about the central axis, and ω its angular momentum. ($I = \frac{1}{2}mR^2$ for cylinders, where R is the radius.)

In the process of rolling down the incline, the cylinder's potential energy turns to kinetic, the total change in each being equal to

$$\Delta E_p = \Delta E_k$$

$$mgh = \frac{1}{2}mv^2 + \frac{1}{2}I\omega^2$$

$$= \frac{1}{2}mv^2 + \frac{1}{2}\left(\frac{1}{2}mR^2\right)\left(\frac{v}{R}\right)^2$$

$$= \frac{1}{2}mv^2 + \frac{1}{4}mv^2 = \frac{3}{4}mv^2$$

using
$$\omega = \frac{v}{R}$$

Thus,
$$gh = \frac{3}{4}v^2$$

$$v = \frac{2\sqrt{3}}{3}(gh)^2 = 1.15\left[\left(9.8 \text{ m/sec}^2\right)\left(2.0 \text{ m}\right)\right]^{1/2}$$

$$= 5.09 \text{ m/sec}$$

Note that the linear speed does not depend upon the size or mass of the cylinder. To find ω, we use the formula

$$\omega = \frac{v}{R} = \frac{5.09 \text{ m/sec}}{0.15 \text{ m}} = 34 \text{ rad/sec}$$

9.7 Conservation of Angular Momentum

When no external force is acting on a system,

$$I_\omega = I_0\omega_0 = \text{constant}$$

angular momentum is conserved.

Problem Solving Example:

Q A man stands at the center of a turntable, holding his arms extended horizontally with a 5 kg weight in each hand. He is set rotating about a vertical axis with an angular velocity of one revolution in 2 seconds. Find his new angular velocity if he drops his hands to his sides. The moment of inertia of the man may be assumed constant and equal to 2 kg-ft². The original distance of the weights from the axis is 7 m and their final distance is 0.1 m.

 If friction in the turnable is neglected, no external torques act about a vertical axis and the angular momentum about this axis is constant. That is,

$$I\omega = (I\omega)_0 = I_0\omega_0$$

where I and ω are the final moment of inertia and angular velocity, and I_0 and ω_0 are the initial values of these quantities.

$$I = I_{man} + I_{weights}$$

The moment of inertia of a weight at a distance r from the axis of rotation is given by

$$I = mr^2$$

Therefore,

$$I = 2 + 2\left(\frac{5}{9.8}\right)\left(\frac{1}{2}\right)^2 = 2.25 \text{ kg-ft}^2$$

$$I_0 = 2 + 2\left(\frac{5}{9.8}\right)(3)^2 = 11.18 \text{ kg-ft}^2$$

$$\omega_0 = 2\pi f_0 = (2\pi)\left(\frac{1}{2} \text{ rev/sec}\right) = \pi \text{ rad/sec}$$

where f_0 is the original frequency of rotation.

$$\omega = \omega_0 \frac{I_0}{I} = 4.97\pi \text{ rad/sec}$$

That is, the angular velocity is more than doubled.

Quiz: Rotational Dynamics

1. On which of the following is the rotational inertia of an object *not* dependent?

    (A) Its axis of rotation.

    (B) Its shape.

    (C) The distribution of mass.

    (D) Its velocity.

    (E) Its direction.

2. A right circular cylinder of radius r rolls down an incline from height h. Determine the ratio of its speed at the bottom to the speed of a point object following the same point.

    (A) 1

    (B) $\sqrt{2}$

    (C) $\sqrt{3}$

    (D) $\sqrt{\dfrac{2}{3}}$

    (E) 2

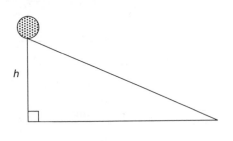

3. The moment of inertia is the rotational analog of

   (A) momentum.

   (B) mass.

   (C) density.

   (D) moment of force.

   (E) None of these.

4. Which statement is true for the mass-pulley system depicted below?

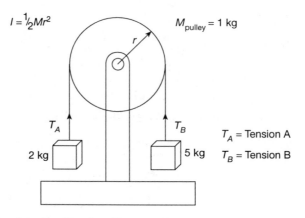

$I = \frac{1}{2}Mr^2$ $M_{pulley} = 1\ kg$

T_A T_B T_A = Tension A

2 kg   5 kg T_B = Tension B

   (A) Tension A > Tension B

   (B) Tension A < Tension B

   (C) Tension A = Tension B

   (D) There is no tension on string A or B.

   (E) The system will remain at rest so the net force on the system is zero.

5. In an Atwood's machine where the one hanging mass is four times the other, find the acceleration.

    (A) $\dfrac{g}{2}$

    (B) $\dfrac{2g}{3}$

    (C) $\dfrac{3g}{5}$

    (D) $\dfrac{3g}{4}$

    (E) $\dfrac{4g}{5}$

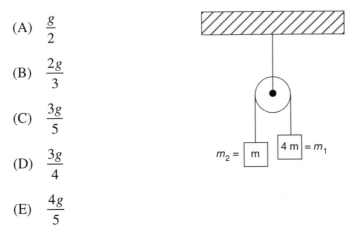

6. A cylinder with a moment of inertia I_0 rotates with angular velocity ω_0. A second cylinder with moment of inertia I_1 initially not rotating drops onto the first cylinder and the two reach the same final angular velocity ω_f. Find ω_f.

    (A) $\omega_f = \omega_0$

    (B) $\omega_f = \dfrac{\omega_0 I_0}{I_1}$

    (C) $\omega_f = \dfrac{I_0 \omega_0}{(I_0 + I_1)}$

    (D) $\omega_f = \dfrac{\omega_0 I_1}{I_0}$

    (E) $\omega_f = \dfrac{\omega_0 (I_0 + I_1)}{I_0}$

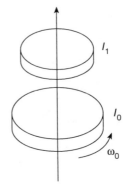

7. If you find yourself sitting atop a Ferris wheel facing north and rotating forward, the direction of the torque according to the right-hand rule is

    (A) south.

    (B) up.

    (C) west.

    (D) east.

    (E) down.

8. If a force of 5 pounds is applied 5 inches from the hinge of a nutcracker, the resistance offered by a nut placed one inch from the hinge is

    (A) 15 lbs.

    (B) 20 lbs.

    (C) 25 lbs.

    (D) 30 lbs.

    (E) 50 lbs.

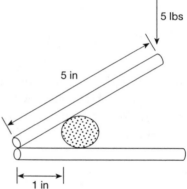

9. Torque is measured in

    (A) pounds.

    (B) joules.

    (C) newtons.

    (D) ft-lbs.

    (E) kilograms.

QUESTION 10 refers to the following diagram.

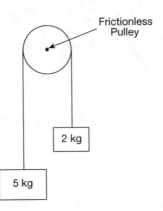

Frictionless
Pulley

2 kg

5 kg

10. The acceleration of the 5 kg mass (in m/s²) is

(A) 4.2.

(D) .428.

(B) 3.92.

(E) .4.

(C) 2.33.

ANSWER KEY

1.	(D)	6.	(C)
2.	(D)	7.	(C)
3.	(B)	8.	(C)
4.	(B)	9.	(D)
5.	(C)	10.	(A)

CHAPTER 10

Harmonic Motion

10.1 Oscillations—Simple Harmonic Motion (SHM)

a) Equations of Motion—Simple Oscillation

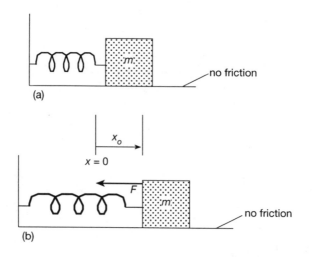

Figure 10.1 Oscillation of a Spring with Mass

$$\overline{F} = -k\overline{x} \;\rightarrow \text{units: newtons}$$

$$\overline{a} = \frac{-k}{m}\,\overline{x} \rightarrow \text{units:} \frac{\text{meters}}{\text{sec}^2}$$

F = restoring force

k = spring constant

x = displacement

a = acceleration

m = mass

b) Equation of Motion — The Variables for SHM

The Period of Motion

$$T = \frac{2\pi}{\omega} = 2\pi\sqrt{\frac{m}{k}} \rightarrow \text{units: seconds}$$

The Frequency of Motion

$$\nu = \frac{1}{T} = \frac{\omega}{2\pi} = \frac{1}{2\pi}\sqrt{\frac{k}{m}} \rightarrow \text{units:} \frac{1}{\text{seconds}}$$

The Angular Frequency of Motion

$$\omega = 2\pi\nu = \frac{2\pi}{T} = \sqrt{\frac{k}{m}} \rightarrow \text{units:} \frac{\text{rads}}{\text{sec}}$$

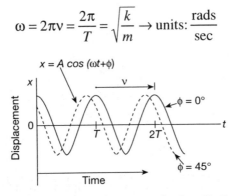

Figure 10.2 Simple Harmonic Oscillation

c) Differential Equations of Motion

Displacement: $\quad \bar{x} = A \cos(\omega t + \phi)$

Velocity: $\quad \dfrac{dx}{dt} = \bar{v} = -\omega A \sin(\omega t + \phi)$

Acceleration: $\quad \dfrac{d^2 x}{dt^2} = \bar{a} = -\omega^2 A \cos(\omega t + \phi)$

x = displacement

v = velocity (linear)

a = acceleration (linear)

A = amplitude

ω = angular velocity

t = time

θ = phase angle

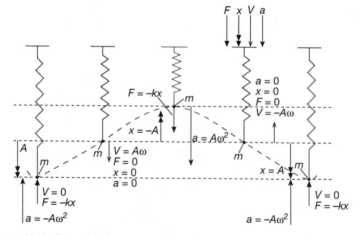

Figure 10.3 Simple Harmonic Oscillation with a Constant Applied Force

For maximum velocity,

$$v_{\max} = \omega A$$

For maximum acceleration,

$$a_{\max} = \omega^2 A$$

Example of displacement equation:

$$x = 0.53 \cos(9.3t + 46)$$

$$\text{amplitude} = 0.53$$

$$\text{angular frequency} = 9.3$$

$$\text{phase angle} = 46$$

Problem Solving Examples:

Q An automobile moves with a constant speed of 50 km/hr around a track of 1 km diameter. What is the angular velocity and the period of the motion?

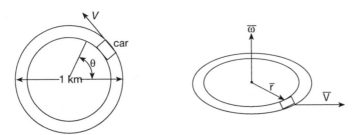

A For circular motion, the angular velocity ω, the radius r, and the linear velocity v obey the relation: $\overline{v} = \overline{\omega} \times \overline{r}$ as shown in the figure. Since ω and r are perpendicular to each other, this reduces to $v = \omega r$.

$$\omega = \frac{v}{r} = \frac{50 \text{ km/hr}}{0.5 \text{ km}} = 100 \text{ rad/hr}$$

The period τ is the time duration of one complete cycle of motion around the circular path. In linear motion, $x = vt$. This equation can be applied to circular motion with linear velocity v replaced by ω and linear distance x replaced by θ expressed in radians. In one cycle of motion, the automobile travels 2π radians.

Therefore, $\omega t = \theta \quad t = \dfrac{\theta}{\omega} \quad \tau = \dfrac{2\pi}{\omega}$

$$\tau = \frac{2\pi}{\omega} = \frac{2\pi}{100 \text{ rad/hr}} = .063 \text{ hr} = 3.8 \text{ min}$$

 Suppose that a mass of 8 grams is attached to a spring that requires a force of 1,000 dynes to extend it to a length 5 cm greater than its natural length. What is the period of the simple harmonic motion of such a system?

 An interesting property of springs is that the length that they stretch is directly proportional to the applied force. The magnitude of this force is

$$F = kx$$

where k is the force constant. The force constant is

$$k = \frac{F}{x} = \frac{1,000 \text{ dynes}}{5 \text{ cm}}$$

$$= 200 \frac{\text{dynes}}{\text{cm}}$$

Therefore, the period is by definition

$$\tau = 2\pi\sqrt{\frac{m}{k}}$$

$$= 2\pi\sqrt{\frac{8 \text{ g}}{200 \text{ dynes/cm}}}$$

$$= 2\pi \sqrt{\frac{8 \text{ g}}{200 \text{ g-cm/cm-sec}^2}}$$

$$= 2\pi \sqrt{\frac{4}{100} \text{ sec}^2}$$

$$= 2\pi \times 0.2 \text{ sec}$$

$$= 1.26 \text{ sec}$$

and the frequency is by definition

$$\nu = \frac{1}{\tau} = \frac{1}{1.26 \text{ sec}} = 0.8 \text{ Hz}$$

10.2 Energy Considerations of SHM

a) Potential Energy, U

$$U = \frac{1}{2} kx^2$$

$$= \frac{1}{2} kA^2 \cos^2(\omega t + \phi)$$

b) Kinetic Energy, K

$$K = \frac{1}{2} mv^2$$

$$= \frac{1}{2} m\omega^2 A^2 \sin^2(\omega t + \phi)$$

$$= \frac{1}{2} kA^2 \sin^2(\omega t + \phi)$$

c) Total Energy, T

$$T = U + K$$

$$T = \frac{1}{2}kA^2$$

d) Velocity, $\overline{v}$

$$\overline{v} = \frac{dx}{dt} = \pm\sqrt{\frac{k}{m}\left(A^2 - x^2\right)}$$

Problem Solving Example:

Q A horizontal shelf moves vertically with simple harmonic motion, the period of which is 1 s and the amplitude of which is 30 cm. A light particle is laid on the shelf when it is at its lowest position. Determine the point at which the particle leaves the shelf and the height to which it rises from that position, g being taken as π^2 m·s^{-2}.

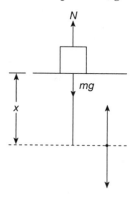

The period of a simple harmonic motion is given by the expression

$$T = \frac{1}{f} = \frac{2\pi}{2\pi f} = \frac{2\pi}{\omega}$$

Therefore, the angular frequency in this case is

$$\omega = \frac{2\pi}{T} = \frac{2\pi}{1s} = 2\pi \ rad \bullet s^{-1}$$

The only forces acting on the particle are its weight $m\,\overline{g}$ downward and the normal force $\overline{N}$ exerted by the shelf upward. (See the figure on previous page.) At any time, according to Newton's second law, $N - mg = ma$, where a is the upward acceleration of the shelf and particle. If $a = -g$ (i.e., the shelf accelerates downward with magnitude g), N becomes zero and, if a becomes more negative, the shelf is retarded at a greater rate than the particle; therefore, the particle moves away from the shelf. Since the shelf undergoes simple harmonic motion, its displacement may be described by $x = A \cos(\omega t + \delta)$, where A is the amplitude of the oscillation, and δ is a phase factor dependent on the initial conditions (i.e., where the shelf is at $t = 0$).

The acceleration

$$a = \frac{d^2x}{dt^2} = -\omega^2 A \cos(\omega t + \delta) = -\omega^2 x$$

Hence, the displacement x at which the particle leaves the shelf is given by

$$-g = -\omega^2 x$$

or

$$x = \frac{g}{\omega^2} = \frac{\pi^2 m \bullet s^{-2}}{4\pi^2 rad^2 \bullet s^{-2}} = \frac{1}{4} m$$

for

$$g = \pi^2 m \bullet s^{-2} \text{ and } \omega = 2\pi \ rad \bullet s^{-1}$$

The particle thus leaves the shelf when it is $1/4$ m above the mean position. At that point the common velocity of shelf and particle is given by the formula relating to velocity to displacement for simple harmonic motion

$$v = \pm \omega \sqrt{A^2 - x^2}$$

Since the shelf is rising, v is positive and

$$v = 2\pi \text{ rad} \bullet \text{s}^{-1} \times \sqrt{(0.3 \text{ m})^2 - (0.25 \text{ m})^2}$$
$$= 2\pi \sqrt{0.0275} \text{ m} \bullet \text{s}^{-1}$$

We now have a new problem concerning a particle thrown upward from a platform with an initial speed v. If the platform level is taken as the reference level for measuring potential energy, then the law of conservation of energy requires that at each moment of time the particle is in motion, the sum of its kinetic and potential energies must remain constant. At the time the particle leaves the platform, its potential energy is zero and

$$E_T = PE + KE = 0 + \frac{1}{2}mv^2$$

At its maximum height h, $v = 0$ and

$$E_T = PE + KE = mgh + 0$$

$\therefore$
$$\frac{1}{2}mv^2 = mgh$$

or
$$h = \frac{v^2}{2g} = \frac{4\pi^2 \times 0.0275 \text{ m}^2 \bullet \text{s}^{-2}}{2 \times \pi^2 \text{ m} \bullet \text{s}^{-2}} = 0.055 \text{ m} = 5.5 \text{ cm}$$

10.3 Pendulums

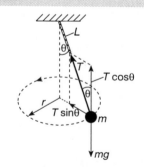

Figure 10.4 The Conical Pendulum

A mass revolves in a horizontal circle of radius r. The velocity is v of constant magnitude. The centripetal force is

$$F = T\sin\theta = m\frac{v^2}{r}.$$

a) Simple Pendulum

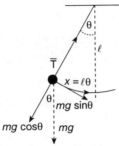

Figure 10.5 Simple Pendulum

i) Force, F

$$F = -mg\theta = \frac{-mg}{L}x$$

ii) Period, T

$$T = 2\pi\sqrt{\frac{L}{g}}$$

L = length of supporting cord

b) Torsional Pendulum

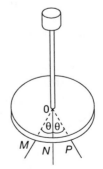

Figure 10.6 Torsional Pendulum

i)   Torque, τ

$$\tau = -k\theta$$

k = torsional constant

ii)   Period, T

$$T = 2\pi\sqrt{\frac{I}{k}}$$

Problem Solving Example:

 A simple pendulum consists of a mass m hung on the end of a string of length L. Find the natural frequency for small oscillations.

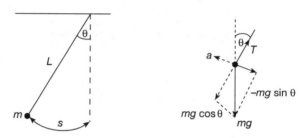

 We start by drawing a diagram of the forces acting on the mass m. The restoring force in the direction of motion is $-mg\sin\theta$.

Thus, the equation of motion is $ma = -mg\sin\theta$

Now we can suppose that θ is small so that we can make the approximation $\sin\theta \cong \theta$. This is accurate to better than 1 percent for $\theta = 15°$ and is better than 5 percent for $\theta = 30°$.

The displacement of the mass is given by the arc s.

$$s = L\theta$$

Thus, $$v = \frac{\Delta s}{\Delta t} = \frac{L\Delta\theta}{\Delta t} = L\omega$$

where

$$\omega = \frac{\Delta\theta}{\Delta t}$$

equals angular velocity. And the acceleration a is given by

$$a = \frac{\Delta v}{\Delta t} = \frac{L\Delta\omega}{\Delta t} = L\alpha$$

where

$$\alpha = \frac{\Delta\omega}{\Delta t}$$

equals angular acceleration. Then finally the equation of motion reduces to

$$mL\alpha = -mg\theta$$

or

$$\alpha + \left(\frac{g}{L}\right)\theta = 0$$

The angular acceleration α is proportional to the negative of the angular displacement θ, so that the motion is simple harmonic with natural angular frequency ω_0 given by

$$\alpha + \omega_0^2\theta = 0$$

where

$$\omega_0^2 = \frac{g}{L}$$

10.4 Simple Harmonic Motion and Uniform Circular Motion

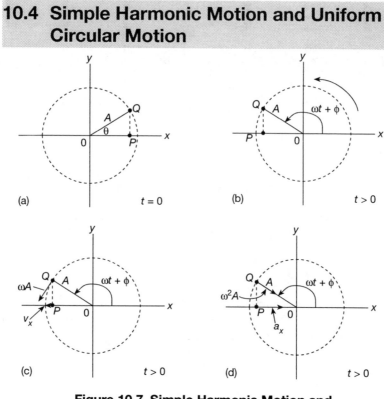

Figure 10.7 Simple Harmonic Motion and Uniform Circular Motion

Equation of Motion

Radius, r

$$r = \sqrt{x^2 + y^2} = A$$

Velocity, v

$$v = \sqrt{v_x^2 + v_y^2} = \omega A$$

Acceleration, a

$$a = \sqrt{a_x^2 + a_y^2} = \omega^2 A$$

Problem Solving Example:

Q A vertical spring has an unstretched length L. When a mass m hangs at rest from its lower end, its length increases to $L + \ell$. Find the period of small vertical oscillations of m (figure).

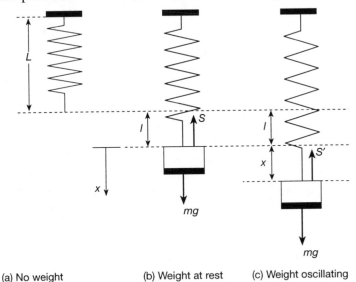

(a) No weight   (b) Weight at rest   (c) Weight oscillating

A When the mass is hanging at rest (Figure (b)), the extension of the spring is ℓ and the force exerted by it on m is, according to Hooke's law

$$S = -k\ell$$

The positive direction is downward, but S is an upward force and therefore negative. Since the mass has no acceleration, this upward force exerted by the spring must be exactly counterbalanced by the weight, mg, and by Newton's second law,

$$F_{net} = -k\ell + mg = 0$$

Hence,

$$mg = k\ell \qquad (1)$$

Suppose that, during the oscillation, the mass is a distance x below its equilibrium position, so that the extension of the spring is $\ell + x$ (Figure (c) on previous page). The force exerted on m by the spring is then

$$S' = -k(\ell + x) = -k\ell - kx$$

The total force F on m is the sum of S' and the weight mg.

$$F = S' + mg$$

$$F = -k\ell - kx + mg$$

According to equation (1) $-k\ell$ cancels $+ mg$, so

$$F = -kx \tag{2}$$

where x is the extension of the spring from its equilibrium position (Figure (c)). In order to find the period of small vertical oscillations of m, we must solve the equation of motion (2) for $x(t)$. Noting that

$$F = ma = m\frac{d^2x}{dt^2}$$

we may write, from (2)

$$m\frac{d^2x}{dt^2} = -kx$$

$$\frac{d^2x}{dt^2} + \left(\frac{k}{m}\right)x = 0 \tag{3}$$

We define

$$\omega_0^2 = \frac{k}{m} \tag{4}$$

Using (3) and (4)

$$\frac{d^2x}{dt^2} + \omega_0^2 x = 0 \tag{5}$$

This is a linear, second order differential equation for x in terms of the variable t. A valid method of solution is to make an educated guess for $x(t)$, substitute this guess into (5), and see if an identity results. If so, $x(t)$ is the solution of (5). A good guess for $x(t)$ is

$$x(t) = A \cos(at + \phi) \tag{6}$$

where a, A, and ϕ are arbitrary constants. Noting that

$$\frac{dx(t)}{dt} = -aA \sin(at + \phi)$$

$$\frac{d^2 x(t)}{dt^2} = -a^2 A \cos(at + \phi)$$

and substituting these results into (5), we obtain

$$-a^2 A \cos(at + \phi) + \omega_0^2 A \cos(at + \phi) = 0$$

which is an identity if $\omega_0^2 = a^2$. Hence, $x(t)$ is a solution if

$$\omega_0 = a$$

and $$x(t) = A \cos(\omega_0 t + \phi)$$

To find the period of motion, note that the cosine function goes through one complete cycle of variation when its argument $(\omega_0 t + \phi)$ has gone through 2π radians. Here the change in the argument is

$$(\omega_0 t_f + \phi) - (\omega_0 t_0 + \phi)$$

or $$\omega_0 (t_f - t_0)$$

and this must equal 2π radians or

$$\omega_0 (t_f - t_0) = 2\pi$$

$$t_f - t_0 = \frac{2\pi}{\omega_0} \tag{7}$$

But $t_f - t_0$ is the time required for cosine to go through one cycle, which is defined as the period of the function (T). Using (7) and (4),

$$T = \frac{2\pi}{\omega_0} = 2\pi\sqrt{\frac{m}{k}} \tag{8}$$

From (1), $\qquad\qquad\qquad mg = k\ell$

or $\qquad\qquad\qquad\qquad k = \frac{mg}{\ell}$

Inserting this in (8), we obtain

$$T = 2\pi\sqrt{\frac{\ell}{g}}$$

Quiz: Harmonic Motion

1. A mass swings freely back and forth in an arc from point A to point D, as shown below. Point B is the lowest point, C is located 1.0 meters above B, and D is 2.0 meters above B. Air resistance is negligible. $g = 10$ m/s^2. The velocity of the mass at point B is closest to

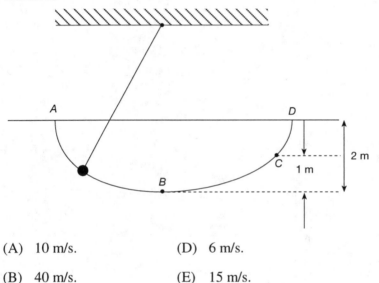

(A) 10 m/s.   (D) 6 m/s.

(B) 40 m/s.   (E) 15 m/s.

(C) 20 m/s.

2. In the diagram to the right, a mass on a spring is pulled down a distance, *−x*, from its equilibrium position, *P*, and released. As the object moves back to *P* and up an additional distance, + *x*, its acceleration graph looks as follows (assume upward acceleration is positive):

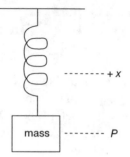

(A)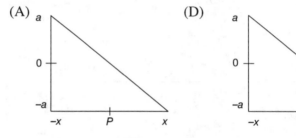

(B)

(C)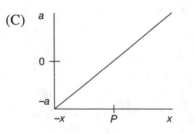

(D)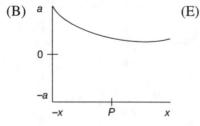

(E)

QUESTIONS 3 and 4 refer to the following figure and information.

I. Y_0 IV. Y_1 and Y_2

II. Y_1 V. Y_0 and Y_1

III. Y_2

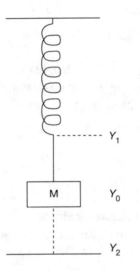

3.   Where is the net force on mass M a minimum?

(A) I   (D) IV

(B) II   (E) V

(C) III

4.   Where is the acceleration of mass M a maximum?

(A) I   (D) IV

(B) II   (E) V

(C) III

5. Which one of the following will cause the period of a pendulum to be doubled?

    (A) Doubling the length

    (B) Doubling the mass

    (C) Doubling the acceleration of gravity

    (D) Increasing the mass by a factor of 4

    (E) Increasing the length by a factor of 4

6. A 1.0 kg pendulum is released at a height of 3.2 m vertically from a reference level. Assuming $g = 10$ m/s^2, and neglecting air resistance, at the bottom of its swing its speed will be

    (A) 8 m/s.   (D) 2 m/s.

    (B) 6 m/s.   (E) 1 m/s.

    (C) 4 m/s.

7. When a 4.0 kg mass is hung vertically on a light spring that obeys Hooke's law, the string stretches 2.0 cm. How much work must an external agent do to stretch the spring 4.0 cm from its equilibrium position?

    (A) 1.57 J   (D) 3.14 J

    (B) 0.39 J   (E) 0.78 J

    (C) 0.20 J

8. Two equal masses $m_1 + m_2 = m$ are connected by a spring having Hooke's constant k. If the equilibrium separation is ℓ_0 and the spring rests on a frictionless horizontal surface, then derive ω the angular frequency.

(A) $\sqrt{\dfrac{k}{m}}$ (D) $2\sqrt{\dfrac{k}{m}}$

(B) $\sqrt{\dfrac{2k}{m}}$ (E) $\sqrt{\dfrac{g}{\ell_0}}$

(C) $\sqrt{\dfrac{3k}{m}}$

9. A pendulum bob of mass m is raised to a height h and released. After hitting a spring of nonlinear force law

$$F = -kx - bx^3,$$

calculate the compression distance x of the spring.

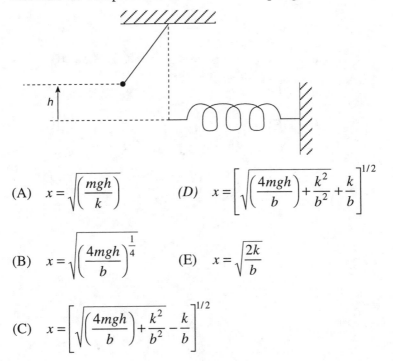

(A) $x = \sqrt{\left(\dfrac{mgh}{k}\right)}$ (D) $x = \left[\sqrt{\left(\dfrac{4mgh}{b}\right) + \dfrac{k^2}{b^2}} + \dfrac{k}{b}\right]^{1/2}$

(B) $x = \sqrt{\left(\dfrac{4mgh}{b}\right)^{\frac{1}{4}}}$ (E) $x = \sqrt{\dfrac{2k}{b}}$

(C) $x = \left[\sqrt{\left(\dfrac{4mgh}{b}\right) + \dfrac{k^2}{b^2}} - \dfrac{k}{b}\right]^{1/2}$

10. The total energy of a particle in SHM is proportional to

(A) the wavelength.

(B) the square of the amplitude.

(C) the phase angle.

(D) the period.

(E) the square of the wavelength.

ANSWER KEY

1.	(D)		6.	(A)
2.	(A)		7.	(A)
3.	(A)		8.	(B)
4.	(D)		9.	(C)
5.	(E)		10.	(B)

CHAPTER 11

Sound Waves

11.1 Speed of Sound

a) Wave Speed in Fluid*

$$v = \sqrt{\frac{B}{\rho_0}} \rightarrow \text{units:} \frac{\text{meters}}{\text{seconds}}$$

v = speed of sound in a fluid

B = modulas of elasticity

ρ_0 = density of medium

*Note: Gas is also a fluid; yet a more precise equation may be given, as seen below.

b) Wave Speed in a Gas

$$v_g = \sqrt{\frac{\gamma p_0}{\rho_0}} \rightarrow \text{units:} \frac{\text{meters}}{\text{seconds}}$$

v_g = speed of sound (Ideal gas)

γ = ratio of specific heats for a gas

p_0 = undisturbed pressure

ρ_0 = density of medium

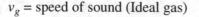

Problem Solving Example:

 Compute the speed of sound in the steel rails of a railroad track. The weight density of steel is 7,854 kg/m³, and Young's modulus for steel is 2.0×10^8 kg/m².

 For an elastic medium, the speed of longitudinal waves is given by

$$v = \sqrt{\frac{Y}{\rho}} = \sqrt{\frac{Yg}{D}}$$

where Y is Young's modulus, ρ is the density of the medium, and $\rho = D/g$ where D is the weight density of the medium.

$$v = \sqrt{\frac{\left(2 \times 10^3 \text{ kg/m}^2\right)\left(9.8 \text{ m}^2\right)}{7,854 \text{ kg/m}^3}}$$

Therefore, $v = 499$ m/s.

11.2   Intensity of Sound

Average Intensity

$$I = \frac{1}{2} \frac{P^2_m}{\sqrt{B\rho_0}}$$

I = average intensity

P_m = pressure amplitude

B = bulk modulus of elasticity

ρ_0 = density of medium

11.3 Allowed Frequencies of a String, Fixed at Both Ends, or an Organ Pipe, Open at Both Ends

$$\nu_n = \frac{n}{2\ell}v, \quad n = 1, 2, 3, \ldots$$

ν_n = frequency

ℓ = length of string or pipe

v = speed of wave

n = number of modes

Problem Solving Example:

What is the lowest frequency of the standing sound wave that can be set up between walls that are separated by 8 m?

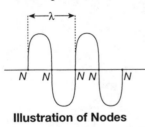

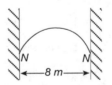

Illustration of Nodes **Lowest Frequency Standing Wave**

A sinusoidal wave that maintains its overall shape between two termination points is called a standing wave. Its amplitude does change with time. For such a wave, the end points have zero amplitude at all times. Zero amplitude points are called nodes and occur every half wavelength.

The wave of lowest frequency has the longest wavelength. For a given distance L, the standing wave of longest wavelength has $\lambda = 2L$ with the only nodes occurring at the termination points. Therefore,

$$\lambda = 2L = 2 \times 8 \text{ m} = 16 \text{ m}$$

The frequency of the wave is

$$\nu = \frac{\upsilon}{\lambda} = \frac{335 \text{ m/sec}}{16 \text{ m}} = 21 \text{ Hz}$$

which is close to the lowest frequency that can be heard by a human ear. Therefore, a room somewhat larger than 25 meters is necessary in order to set up standing waves of the lowest audible frequency (for example, organ notes of 16 Hz).

11.4 The Beat Equation for Particle Displacement

$$y = \left[2y_m \cos 2\pi \left(\frac{\nu_1 - \nu_2}{2} \right) t \right] \cos 2\pi \left(\frac{\nu_1 + \nu_2}{2} \right) t$$

y = displacement

y_m = amplitude

ν_1 = frequency of wave 1

ν_2 = frequency of wave 2

t = time

11.5 The Beat Frequency

$$\nu_B = \nu_1 - \nu_2$$

where ν_B = beat frequency

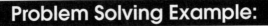

Problem Solving Example:

 When two tuning forks are sounded simultaneously, a beat note of 5 cycles per second is heard. If one of the forks has a known frequency of 256 cycles per second, and if a small piece of adhesive tape fastened to this fork reduces the beat note to 3 cycles per second, what is the frequency of the other fork?

 This problem involves the phenomenon of beats. When two similar waves are superimposed, the beat frequency represents the numerical difference in their frequencies. Hence, for the case in question

$$n = (256 \pm 5) \text{ cycles/sec}$$

where n represents the unknown frequency.

It appears that n has two possible values, either 251 or 261. Now, when the standard fork is loaded with the tape, its frequency will decrease. Since the beat frequency is then reduced to 3 cycles per second, the unknown frequency must be less rather than more than 256. Hence,

$$n = 251.$$

11.6 The Doppler Effect

a) Source at Rest

$$v' = v \left(\frac{v \pm v_0}{v} \right)$$

v' = observed (heard) frequency

v = actual frequency

v = speed of sound in a medium

v_0 = speed of observer

Note: The $\pm$ signs denote when the motion of sound is toward or away from the source; it's $+$ when the motion is toward the source, $-$ when the motion is away from the source.

b) Observer at Rest

$$v' = v\left(\frac{v}{v \pm v_s}\right)$$

where v_s is the speed of the source.

c) The General Doppler Equation:

$$v' = v\left(\frac{v \pm v_0}{v \pm v_s}\right)$$

Problem Solving Examples:

Q Two trains moving along parallel tracks in opposite directions converge on a stationary observer, each sounding its whistle of frequency 350 cycles $\bullet$ s^{-1}. One train is traveling at 50 kph. What must be the speed of the other if the observer hears five beats per second. The speed of sound in air is 750 kph.

A When a source is moving toward a stationary observer, the latter hears a frequency for the emitted note which is related to the frequency of the source by the expression $f = uf_s / (u - v_s)$, where u is the speed of sound in air, v_s is the speed of the source, and f_s is the frequency of the sound emitted by the source. If one of the trains is moving toward the observer with v_1 and the other with speed v_2, then since both have the same frequency whistle,

$$f_1 = \frac{uf_s}{u - v_1}$$

and

$$f_2 = \frac{uf_s}{u - v_2} = \frac{750 \text{ kph} \times 350 \text{ s}^{-1}}{(750 - 50) \text{ kph}}$$

$$= 375 \text{ cycles} \bullet \text{s}^{-1}$$

But the observer hears five beats per second. This corresponds to a frequency difference

$$f_1 - f_2 = \pm 5 \text{ cycles} \bullet \text{s}^{-1}$$

Hence, $\qquad\qquad f_1 = 370 \text{ cycles} \bullet \text{s}^{-1}$

or $\qquad\qquad\qquad 380 \text{ cycles} \bullet \text{s}^{-1}$

$$\therefore f_1 = \frac{uf_s}{u - v_1} = 370 \text{ s}^{-1} \quad \text{or} \quad 380 \text{ s}^{-1}$$

$$\therefore u - v_1 = \frac{750 \text{ kph} \times 350 \text{ s}^{-1}}{370 \text{ s}^{-1}} \quad \text{or} \quad \frac{750 \text{ kph} \times 350 \text{ s}^{-1}}{380 \text{ s}^{-1}}$$

$$= 709.5 \text{ kph} \qquad \text{or} \quad 690.8 \text{ kph}$$

$$\therefore v_1 = 40.5 \text{ kph} \qquad \text{or} \quad 59.2 \text{ kph}$$

Q A researcher notices that the frequency of a note emitted by an automobile horn appears to drop from 284 cycles $\bullet$ s^{-1} to 266 cycles $\bullet$ s^{-1} as the automobile passes him. From this observation he is able to calculate the speed of the car, knowing that the speed of sound in air is 335 m $\bullet$ s^{-1}. What value does he obtain for the speed?

A This is an example illustrating the Doppler effect. When there is no movement of the surrounding medium, the relation between the frequency as heard by a moving observer and that emitted by a moving source is

$$\frac{f_L}{u \pm v_L} = \frac{f_s}{u \pm v_s}$$

where f_L is the frequency heard by the listener, f_s the frequency emitted by the moving source, v_L the velocity of the listener, v_s the velocity of the source, and u the velocity of sound ($= 335$ m $\bullet$ s^{-1}). The upper signs (+ left side of equation, − right side) correspond to the source and observer moving along the line joining the two and approaching each other and the lower signs (− left, + right) correspond to source and observer receding from one another.

In this case the frequencies heard by the stationary listener ($v_L = 0$) will be $f_L + uf_s / (u \pm v_s)$. As the automobile approaches the observer, he records a frequency of 284 cycles • s^{-1}, and as the automobile moves away from him, he records 266 cycles • s^{-1}. Thus,

$$284 \text{ s}^{-1} = \frac{uf_s}{u - v_s} \qquad (1)$$

and

$$266 \text{ s}^{-1} = \frac{uf_s}{u + v_s} \qquad (2)$$

Dividing equation (1) by (2)

$$\frac{u + v_s}{u - v_s} = \frac{284}{266}$$

$$266(u + v_s) = 284(u - v_s)$$

$$(266 + 284)v_s = (284 + 266)u$$

$$\frac{v_s}{u} = \frac{18}{550}$$

or

$$\therefore \frac{v_s}{u} = \frac{18}{550} \times 335 \text{ m} \bullet \text{s}^{-1} = 10.9 \text{ m} \bullet \text{s}^{-1} = 39.5 \text{ km/h}$$

11.7 Half Angle of a Shock Wave

$$\sin \theta = \frac{v}{v_s}$$

θ = half angle of mach cone

v = speed of sound in a medium

v_s = speed of source

Quiz: Sound Waves

QUESTIONS 1 and 2 refer to the following:

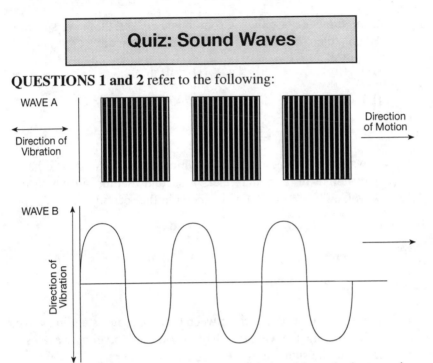

WAVE A

Direction of Vibration

Direction of Motion

WAVE B

Direction of Vibration

1. If the wavelength of wave *B* is doubled and its velocity remains constant, its period *T* will

    (A) remain constant.

    (B) double.

    (C) decrease by $\frac{1}{2}$.

    (D) increase by four times.

    (E) None of the above.

2. A sound wave traveling through air is best described by wave *A*. The dark regions in the wave are called

    (A) rarefactions.

    (B) compressions.

    (C) interference.

    (D) sound bands.

    (E) ripples.

3. If the frequency of a wave source is 5 Hz and the wave speed is 10 m/s, then the distance between successive wave crests is

    (A) 1 m.   (D) 10 m.

    (B) 2 m.   (E) 50 m.

    (C) 5 m.

4. A wave leaves one medium in which it travels 6 m/s and has a wavelength of 2 m and enters a second medium in which the wavelength is 3 m. The wave speed in the second wave is

    (A) 1.5 m/s.   (D) 9 m/s.

    (B) 5 m/s.   (E) 18 m/s.

    (C) 6 m/s.

5. If a segment, the distance between two nodes in a standing wave, is 20 cm long, the wavelength of the original traveling wave used to create the standing wave is

    (A) 10 cm.   (D) 40 cm.

    (B) 20 cm.   (E) 50 cm.

    (C) 30 cm.

6. Two sounds close in frequency, f_1 and f_2, will produce a third sound (or beating) having a frequency determined by

    (A) $f_2 - f_1$.   (D) $\dfrac{f_1}{f_2}$.

    (B) $f_1 + f_2$.   (E) f_1, the lower pitched sound.

    (C) $f_1 \times f_2$.

7. When the frequency of a forced vibration approaches the natural frequency of a particular vibrating object, the vibration in the object increases its

(A) pitch.

(D) speed.

(B) frequency.

(E) amplitude.

(C) wavelength.

8. As a train sounds its horn and moves at a constant speed toward a stationary listener, the listener will hear

(A) the pitch gradually get higher.

(B) a constantly higher pitch.

(C) the same pitch as the engineer.

(D) a constantly lower pitch.

(E) the pitch gradually getting lower.

9. A transverse wave

(A) displaces the medium perpendicular to the wave velocity.

(B) displaces the medium parallel to the wave velocity.

(C) transmits sound.

(D) Both (A) and (C).

(E) Both (B) and (C).

10. The distance from a compression in a sound wave to the nearby rarefaction is 5 m. The frequency of the sound is 33 hertz. The temperature is most near

    (A) 0 degrees K.

    (B) 135 degrees K.

    (C) 273 degrees K.

    (D) 548 degrees K.

    (E) Temperature is not related to the speed of sound.

	ANSWER KEY		
1.	(B)	6.	(A)
2.	(B)	7.	(E)
3.	(B)	8.	(B)
4.	(D)	9.	(A)
5.	(D)	10.	(C)

CHAPTER 12

Gravitation

12.1 Newton's Law of Gravitation

$$F = G\frac{m_1 m_2}{r^2} \rightarrow \text{units: Newtons}$$

F = force of gravity

G = gravitational constant

m = mass

r = distance between m_1 and m_2

$G = 6.6726 \times 10^{-11} \text{ m}^3/\text{kg} \cdot \text{s}^2$

Problem Solving Example:

Compute the force of gravitational attraction between the large and small spheres of a Cavendish balance, if m = 1 gm, m' = 500 gm, and r = 5 cm.

Two uniform spheres attract each other as if the mass of each were concentrated at its center. By Newton's Law of Universal

Gravitation, the force of attraction between two masses m and m' separated by a distance r is

$$F = \frac{Gmm'}{r^2} = \frac{\left(6.67 \times 10^{-8}\,\text{dyne} \bullet \text{cm}^2/\text{gm}^2\right) \times \left(1\,\text{gm}\right) \times \left(500\,\text{gm}\right)}{\left(5\,\text{cm}\right)^2}$$

$$= 1.33 \times 10^{-6}\,\text{dyne}$$

or about one-millionth of a dyne.

12.2 The Motion of Planets and Satellites— Kepler's Laws of Planetary Motion

a) Law of Orbits–All planets move in elliptical orbits having the sun as one focus.

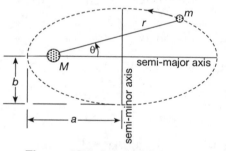

Figure 12.1 Law of Orbits

Equation of Orbit

$$\frac{1}{r} = \frac{1}{b^2}\left(a - \sqrt{a^2 - b^2}\,\cos\theta\right)$$

r = distance between masses

b = semi-minor axis

a = semi-major axis

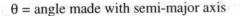

θ = angle made with semi-major axis

b) Law of Areas—A line joining any planet to the sun sweeps out equal areas in equal times.

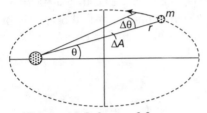

Figure 12.2 Law of Areas

Rate of Area Change

$$\frac{dA}{dt} = \frac{1}{2}r^2\omega$$

A = area

r = distance

ω = angular velocity

c) Law of Periods

$$T^2 = \left(\frac{4\pi^2}{GM}\right)a^3$$

T = period

G = gravitational constant

M = mass

a = semi-major axis

Problem Solving Example:

 Find the period of a communications satellite in a circular or bit 22,300 km above the earth's surface, given that the radius of the earth is 4,000 km, that the period of the moon is 27.3 days, and that the orbit of the moon is almost circular with a radius of 239,000 km.

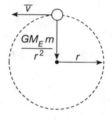

 In a circular orbit of radius r, an earth satellite of mass m has a velocity v. The distance, d, the satellite moves in one revolution equals $2\pi r$. Its period T is the time it takes for the satellite to make one revolution. Therefore,

$$T = \frac{d}{v} = \frac{2\pi r}{v} \qquad (1)$$

The centripetal force necessary to keep the satellite moving in a circle is supplied by the gravitational force exerted by the earth. By Newton's Second Law,

$$G\frac{M_E m}{r^2} = \frac{mv^2}{r}$$

where G is the gravitational constant and M_E is the mass of the earth. From equation (1)

$$\frac{v^2}{r^2} = \left(\frac{2\pi}{T}\right)^2 = \frac{4\pi^2}{T^2}$$

and $\qquad G\frac{M_E m}{r^2} = \frac{4\pi^2 mr}{T^2} \qquad (2)$

The same arguments apply to the moon of mass M_M which moves in a circle of radius R with period T_M.

$$G\frac{M_E\,M_M}{R^2} = \frac{4\pi^2 M_M R}{T_M^2} \tag{3}$$

Solving for GM_E in equations (2) and (3), we have respectively

$$GM_E = \frac{4\pi^2 r^3}{T^2}$$

$$GM_E = \frac{4\pi^2 R^3}{T_M^2}$$

Equating the above two expressions,

$$\frac{4\pi^2 r^3}{T^2} = \frac{4\pi^2 R^3}{T_M^2}$$

or

$$\frac{T^2}{T_M^2} = \frac{r^3}{R^3}$$

Therefore, the period of the satellite can be found by substitution of the numerical values given. Using

$$r = 22{,}300 \text{ km} + R_E = 22{,}300 \text{ km} + 4{,}000 \text{ km} = 26{,}300 \text{ km}$$

the period is

$$T = T_M\sqrt{\frac{r^3}{R^3}} = 27.3 \text{ days} \sqrt{\frac{\left(2.63\times10^4\right)^3 \text{ mi}^3}{\left(2.39\times10^5\right)^3 \text{ mi}^3}}$$

$$= 1.00 \text{ day}$$

Such a satellite therefore rotates about the center of the earth with the same period as the earth rotates about its axis. In other words, if it is rotating in the equatorial plane, it is always vertically above the same point on the earth's surface.

12.3 Angular Momentum of Orbiting Planets

a) Angular Momentum

$$\ell = mr^2\omega$$

ℓ = angular momentum

m = mass

ω = angular velocity

b) Rate of Area Change

$$\frac{dA}{dt} = \frac{\ell}{2m}$$

A = area

ℓ = angular momentum

m = mass

Problem Solving Example:

Q Consider a satellite in a circular orbit concentric and coplanar with the equator of the earth. At what radius r of the orbit will the satellite appear to remain stationary when viewed by observers fixed on the earth? We suppose the sense of rotation of the orbit is the same as that of the earth.

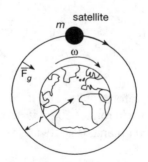

A The satellite is being pulled towards the earth by a gravitational force $\overline{F}_g$. By Newton's second law, this force equals ma, where m is the mass of the satellite and a is the linear acceleration along the direction parallel to the force $\overline{F}_g$ (which acts as the centripetal force in this case since the motion is circular). Therefore,

$$\overline{F}_g = m\overline{a}$$

Furthermore, we know that in circular motion, linear acceleration is given by $a = \omega^2 r$, where ω is the angular velocity of m. The force of gravity ($\overline{F}_g$) is given by GmM/r^2, where G is the gravitational constant and M is the mass of the earth. Substituting in the equation above, we have

$$\frac{GmM}{r^2} = m\omega^2 r$$

Solving for r we have

$$r^3 = \frac{GM}{\omega^2}$$

In this equation we note that G is a constant as is M (the mass of the earth). Therefore, r varies as a function of ω (or vice versa). If we fix ω, we necessarily fix r. For the satellite to appear to remain stationary when viewed by observers fixed on the earth, the satellite must have the same angular velocity as the earth. The angular velocity of the earth ($\omega_e = \Delta\theta/\Delta t$, or change of angle per unit time) is

$$\frac{2\pi}{1\,day} = \frac{2\pi}{(60 \text{ sec/min})(60 \text{ min/hr})(24 \text{ hr/day})}$$

$$= \frac{2\pi}{8.64 \times 10^4 \text{ sec}^{-1}} = 7.3 \times 10^{-5} \text{ sec}^{-1}$$

Substituting in the above equation, we have

$$r^3 = \frac{\left(6.67 \times 10^{-11} \text{N} \cdot \text{m}^2/\text{kg}^2\right)\left(5.98 \times 10^{24} \text{kg}\right)}{\left(7.3 \times 10^{-5} \text{sec}^{-1}\right)^2}$$

$$r^3 \approx 7.49 \times 10^{22} \text{m}^3$$

$$r \approx 4.2 \times 10^7 \text{m}$$

$$r \approx 4.2 \times 10^9 \text{cm}$$

The radius of the earth is 6.38×10^8 cm. The distance calculated is roughly one-tenth of the distance to the moon.

12.4 Gravitational Potential Energy, *U*

$$U = - W_\infty$$

equals work done by gravity on a particle as it moves from infinity to a distance *r* from the earth's center.

$$U = \frac{-GMm}{r}$$

G = gravitational constant

M = mass of denser particle

m = mass of orbiting particle

r = distance

Problem Solving Example:

A 200-kg satellite is lifted to an orbit of 2.20×10^4 km radius. How much additional potential energy does it acquire relative to the surface of the earth?

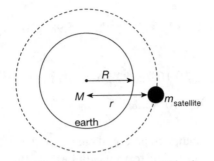

As in the diagram, R is the earth's radius, M is the earth's mass, m is the satellite mass, and r is the distance between the earth's center and the satellite.

$$R = 6.37 \times 10^6 \text{ m}$$

$$r = 2.20 \times 10^4 \text{ km}$$

$$M = 5.98 \times 10^{24} \text{ kg}$$

$$m = 200 \text{ kg}$$

The additional potential energy is to the work done against the earth's gravitational field. At a distance R from the earth's center, that is, on the earth's surface, the satellite has a potential energy, U_{surface}.

$$U_{\text{surface}} = \frac{GMm}{R}$$

In orbit of radius r the potential is

$$U_{\text{orbit}} = -\frac{GMm}{R} .$$

Then the additional potential energy involved in launching the rocket to its orbit, ΔU, is given by

$$\Delta U = U_{\text{orbit}} - U_{\text{orbit}} = -\frac{GMm}{r} - \left(-\frac{GMm}{R} \right) = GMm\left(\frac{1}{R} - \frac{1}{r} \right)$$

$$= \left(6.67 \times 10^{-11} \text{N-m}^2/\text{kg}^2\right)\left(5.98 \times 10^{24} \text{kg}\right)(200 \text{ kg})$$

$$\times \left(\frac{1}{6.37 \times 10^6 \text{m}} - \frac{1}{2.20 \times 10^7 \text{m}}\right)$$

$$= 8.90 \times 10^9 \text{ joules}$$

Note that the change in potential energy of the satellite cannot be found by using $U = mgh$. This formula applies only to objects near the earth's surface, where g is approximately constant.

12.5 Potential Energy for a System of Particles

For three particles:

$$U = -\left(\frac{Gm_1 m_2}{r_{12}} + \frac{Gm_1 m_3}{r_{13}} + \frac{Gm_2 m_3}{r_{23}}\right)$$

For four particles, there are six different gravitational potentials.

For five, there are ten.

For n particles, there are $\frac{1}{2}(n-1)n$ individual gravitational potentials.

Problem Solving Example:

 Let us estimate the gravitational energy of the galaxy. We omit from the calculation the gravitational self-energy of the individual stars.

 The gravitational energy of an arbitrary system of N particles consists of the sum of the mutual potential energies of each pair of particles. Hence,

$$U = \frac{1}{2}(U_{12} + U_{13} + U_{14} + ... + U_{1N}) + \frac{1}{2}(U_{21} + U_{23} + U_{24}$$

$$+ ... + U_{2N}) + (U_{31} + U_{32} + U_{34} + ... + U_{3N} + ...$$

$$+ (U_{N1} + U_{N2} + U_{N3} + ... + U_{NN-1}) \tag{1}$$

The terms U_{12}, etc., represent the mutual potential energy of particles 1 and 2. By including sets of terms such as U_{12} and U_{21}, we have double counted, because these represent the mutual potential energies of particles 1 and 2, and particles 2 and 1, respectively. However, these two terms are the same. Hence, the factor $^{1}/_{2}$ must be included in (1) to negate the process of double counting. Furthermore, terms such as U_{11} are omitted because they represent the mutual potential energy of particle 1 with particle 1, i.e., they are self-energies.

We approximate the gross composition of the galaxy by N stars, each of mass M, and with each pair of stars at a mutual separation of the order of R.

From the definition of potential energy

$$U_{ij} = \frac{-Gm_i m_j}{r_{ij}}$$

where m_i and m_j are the masses of particles i and j, respectively, $G = 6.67 \times 10^{-11}$ N • m^2/kg^2, is their mutual separation. For our case,

$$r_{ij} = R$$

and
$$m_i = m_j = M$$

for all pairs of particles. Then

$$U_{ij} = \frac{-GM^2}{R} \tag{2}$$

for any two particles.

Notice that in equation (1), the first parenthesis has $N - 1$ terms of the type (2), the second parenthesis has $N - 1$ terms of this type, and similarly for all parentheses. Altogether, there are $N(N - 1)$ terms of type (2) in (1). Therefore,

$$U = \frac{1}{2}(N)(N-1)\left(\frac{-GM^2}{R}\right)$$

$$U = -\frac{1}{2}\frac{(N)(N-1)GM^2}{R}$$

If $N \approx 1.6 \times 10^{11}$, $R \approx 10^{21}$, and $M \approx 2 \times 10^{30}$ kg,

$$U \approx \frac{-\frac{1}{2}\left(1.6 \times 10^{11}\right)\left(1.6 \times 10^{11} - 1\right)\left(6.67 \times 10^{-11}\,\text{N} \bullet \text{m}^2/\text{kg}^2\right)\left(2 \times 10^{30}\,\text{kg}\right)^2}{10^{21}\,\text{m}}$$

$$U \approx \frac{-34.15 \times 10^{71}}{10^{21}}\,\text{J} = -34.15 \times 10^{50}\,\text{J}$$

$$U \approx -3.42 \times 10^{51}\,\text{J}$$

12.6 Kinetic Energy

$$K = \frac{1}{2}m\omega^2 r^2{}_0$$

$$K = \frac{1}{2}\frac{GMm}{r_0}$$

Problem Solving Example:

Q Use the principle of conservation of mechanical energy to find the velocity with which a body must be projected vertically upward, in the absence of air resistance, to rise to a height above the earth's surface equal to the earth's radius, R.

Let the center of the earth be the origin. Then the initial distance of the body is $r_1 = R$ and its final position is $r_2 = 2R$. Let v_1 be the initial velocity. v_2, the final velocity of the body of mass, m, is zero since $2R$ is the maximum height the body rises. Using conservation of energy, we have

$$KE_1 + PE_1 = KE_2 + PE_2$$

$$\frac{1}{2}mv_1^2 - G\frac{mm_E}{r_1} = \frac{1}{2}mv_2^2 - G\frac{mm_E}{r_2}$$

where m_E is the earth's mass.

Note that potential energy is negative. Substitution yields

$$\frac{1}{2}mv_1^2 - G\frac{mm_E}{R} = 0 - G\frac{mm_E}{2R}$$

$$v_1^2 = \frac{Gm_E}{R}$$

12.7 Total Energy

$$T = K + U$$

$$T = \frac{1}{2}\frac{GMm}{r_0} - \frac{GMm}{r_0}$$

$$T = -\frac{GMm}{2r_0}$$

Problem Solving Example:

Estimate the average temperature of the interior of the sun. The gravitational self-energy, U_s, of a uniform star of mass M_s and radius R_s is

$$U_s = \frac{1}{n}\frac{3GM_s^2}{5R_s}$$

The average kinetic energy of a single atom in a star is proportional to the absolute temperature T.

$$<\text{K.E. of a particle}> = \frac{3}{2} kT$$

with the constant k (the Boltzmann constant) given by

$$k = 1.38 \times 10^{-16} \text{ erg/deg Kelvin}$$

Here, the brackets $<\ >$ denote an average value.

The total kinetic energy in the star is $3/2\ kNT_{av}$, where T_{av} is an appropriate average temperature over the interior of the star, and N is the number of atoms in the star. Then, the virial theorem gives

$$<\text{K.E. of all atoms}> = -\frac{1}{2} <\text{P.E. of sun}>$$

where P.E. is potential energy. Hence,

$$\frac{3}{2} NkT_{av} = -\frac{3GM_s^2}{10R_s}$$

Thus, we have

$$T_{av} = \frac{GM_s^2}{5R_s Nk} = \frac{GM_s M}{5R_s k} \tag{1}$$

where $M = M_s/N$ is the average mass of an atom in the star. (Most of the atoms in a star are generally hydrogen or helium.)

The mass of the sun, M_s, approximately equals 2×10^{33} gm, and its radius, R_s, is approximately 7×10^{10} cm. Let us take M as 3×10^{-24} gm, about twice the proton mass. Then (1) becomes

$$T_{av} \approx \frac{\left(7 \times 10^{-8} \text{dynes} \bullet \text{cm}^2/\text{gm}^2\right)\left(2 \times 10^{33} \text{gm}\right)\left(3 \times 10^{-24} \text{gm}\right)}{5\left(7 \times 10^{10} \text{cm}\right)\left(1 \times 10^{-16} \text{erg/K}\right)}$$

$$T_{av} \approx 10^7 \text{ K}$$

We have performed what is known as an order of magnitude calculation for T_{av}.

Quiz: Gravitation

1. Two charges are separated by 2.0 m. The force of attraction between them is 4 N. If the distance between them is doubled, the new force between them is

   (A) .5 N.   (D) 4 N.

   (B) 1 N.   (E) 8 N.

   (C) 2 N.

2. Two unequal masses falling freely from the same point above the earth's surface would experience the same

   (A) acceleration.

   (B) decrease in potential energy.

   (C) increase in kinetic energy.

   (D) increase in momentum.

   (E) change in mass.

3. Two objects of equal mass are a fixed distance apart. If the mass of each object could be tripled, the gravitational force between the objects would

   (A) decrease by $\frac{1}{3}$.   (D) increase by nine times.

   (B) decrease by $\frac{1}{9}$.   (E) decrease by nine times.

   (C) triple.

4. Compared to the force exerted on B at a separation of 12 meters, the force exerted on sphere B at a separation of 6 meters would be

    (A) $\dfrac{1}{2}$ as great.

    (D) 4 times as great.

    (B) 2 times as great.

    (E) 9 times as great.

    (C) $\dfrac{1}{4}$ as great.

5. The moon is in a nearly circular orbit above the earth's atmosphere. Which statement is true?

    (A) It is in equilibrium and has no net force.

    (B) It has constant velocity.

    (C) It continues to use up its energy rapidly like a spaceship and is falling back to earth.

    (D) It is accelerating toward the earth.

    (E) Its acceleration is in the same direction as its velocity.

6. A mass is suspended on a spring. The reaction force to the force of gravity from the earth acting on the mass is the force exerted by the

    (A) mass on the earth.

    (D) spring on the earth.

    (B) mass on the spring.

    (E) earth on the mass.

    (C) spring on the mass.

7. An object weighing 100 N at the earth's surface is moved to a distance of 3 earth radii from the surface of the earth. Its new weight will be

(A) 25 N.

(D) 6.25 N.

(B) 33.3 N.

(E) 400 N.

(C) 11.1 N

8. An object orbits a star in an elliptical orbit. The distance at aphelion is $2a$ and the distance at perihelion is a. Determine the ratio of the object's speed at perihelion to that at aphelion.

(A) 2

(B) 3

(C) 1

(D) $\sqrt{2}$

(E) $\sqrt{3}$

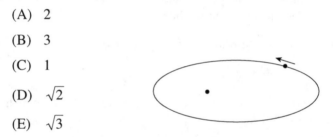

9. What is the gravitational field of an infinite line mass of linear mass density λ?

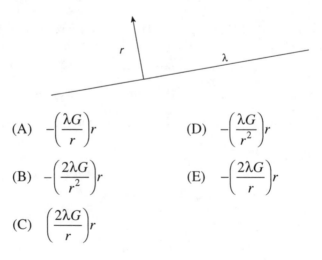

(A) $-\left(\dfrac{\lambda G}{r}\right)r$

(D) $-\left(\dfrac{\lambda G}{r^2}\right)r$

(B) $-\left(\dfrac{2\lambda G}{r^2}\right)r$

(E) $-\left(\dfrac{2\lambda G}{r}\right)r$

(C) $\left(\dfrac{2\lambda G}{r}\right)r$

10. It is possible that Newtonian theory of gravitation may need to be modified at short range. Suppose that the potential energy between two mass m and m' is given by

$$V(r) = -\frac{Gmm'}{r}(1 - ae^{-r/\lambda})$$

For short distances $r<\lambda$, calculate the force between m and m'.

(A) $F = \dfrac{-Gmm'}{r^2}$

(D) $F = \dfrac{-Gmm'\,a}{\lambda r}$

(B) $F = \dfrac{-Gmm'(1-a)}{r^2}$

(E) $F = \dfrac{Gmm'(1-a)}{\lambda r}$

(C) $F = \dfrac{-Gmm'(1+a)}{r^2}$

ANSWER KEY

1.	(B)	6.	(A)
2.	(A)	7.	(D)
3.	(D)	8.	(A)
4.	(D)	9.	(E)
5.	(D)	10.	(B)

CHAPTER 13

Equilibrium of Rigid Bodies

13.1 Rigid Bodies in Static Equilibrium

a) Sum of the Forces

$$\sum_n \overline{F} = \left(\overline{F}_1 + \overline{F}_2 + \ldots + \overline{F}_n \right) = 0$$

b) Sum of the Torques

$$\sum_n \overline{\tau} = \left(\overline{\tau}_1 + \overline{\tau}_2 + \ldots + \overline{\tau}_n \right) = 0$$

Problem Solving Example:

  A light horizontal bar is 4.0 m long. A 3.0-kg force acts vertically upward on it 1.0 m from the right-hand end. Find the torque about each end.

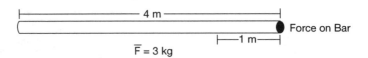

Since the force is perpendicular to the bar, the moment arms are measured along the bar.

About the right-hand end

$$L_r = 3.0 \text{ kg} \times 1.0 \text{ m} = 3.0 \text{ kg-m clockwise}$$

About the left-hand end

$$L_1 = 3.0 \text{ kg} \times 3.0 \text{ m} = 9.0 \text{ kg-m counterclockwise}$$

The torques produced by this single force about the two axes differ in both magnitude and direction. This causes the bar to twist through an angle θ which is proportional to the torque.

13.2 Free-body Diagrams

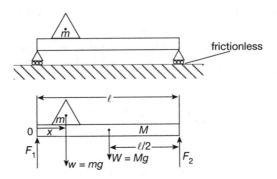

Figure 13.1 Free-body Diagram

$$\Sigma F_x = 0$$

$$\Sigma Fy = F_1 - w - W + F_2 = 0$$

$$\Sigma \tau_0 = F_1(0) - w(x) - W\left(\frac{\ell}{2}\right) + F_2(\ell) = 0$$

Note: In choosing a reference point about which all torques are measured, one may choose any point in the system.

Problem Solving Example:

Q A 100 kg man hangs from the middle of a tightly stretched rope so that the angle between the rope and the horizontal direction is 5°, as shown in Figure (a). Calculate the tension in the rope. (See Figure (b).)

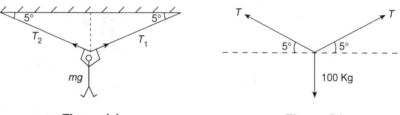

Figure (a)   **Figure (b)**

A Since the two sections of the rope are symmetrical with respect to the man, the tensions in them must have the same magnitude (Figure (b)). This can be arrived at by summing the forces in the horizontal direction and setting them equal to zero since the system is in equilibrium. Then

$$\sum F_x = T_1 \cos 5° - T_2 \cos 5° = 0$$

and $$T_1 = T_2 = T$$

Considering the forces in the vertical direction,

$$\sum F_y = T \sin 5° + T \sin 5° - 100 \text{ kg} = 0$$
$$100 \text{ kg} = 2T \sin 5° = 2T(0.0871)$$
$$T = \frac{(100)}{(2)(0.0871)} = 575 \text{ kg}$$

Note the significant force that can be exerted on objects at either end of the rope by this arrangement. The tension in the rope is over five times the weight of the man. Had the angle been as small as 1°, the tension

would have been

$$T = \frac{100}{2\sin 1°} = \frac{100}{(2)(0.0174)} = 2,865 \text{ kg}$$

This technique for exerting a large force would only be useful to move something a very small distance, since any motion of one end of the rope would change the small angle considerably and the tension would decrease accordingly.

Quiz: Equilibrium of Rigid Bodies

QUESTION 1 refers to the following diagram.

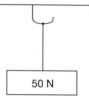

1. A uniform rope of weight 100 newtons hangs from a hook. A box of 50 newtons hangs from the rope. What is the tension in the rope?

    (A) 50 N throughout the rope

    (B) 75 N throughout the rope

    (C) 100 N throughout the rope

    (D) 150 N throughout the rope

    (E) It varies from 100 N at the bottom of the rope to 150 N at the top.

2. The baseball bat pictured is balanced at point x. This implies that

    (A) side L weighs the same as side R.

    (B) side L has a mass that is the same as the mass R.

    (C) the torque of side L equals the torque of side R.

    (D) the density of the wood on side L is less than the density of side R.

    (E) None of the above.

3. The nonuniform bar shown below is 5 meters long. It is denser near one end than near the other. It has a fixed pivot at 2.3 m from the heavy end. If the weight of the bar is 73.3 newtons and a 20 newton weight at 0.5 m from the light end will establish rotational equilibrium, what is the location of the bar's center of mass?

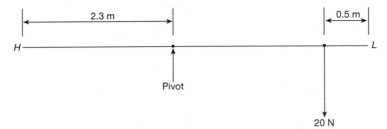

(A) 0.13 m left of pivot

(B) 0.6 m from the heavy end

(C) 1.7 m from the heavy end

(D) 2.5 m from the light end

(E) 2.9 m from the heavy end

4. A normal force is

(A) perpendicular to the Earth.

(B) parallel to the surface an object rests upon.

(C) perpendicular to the surface an object rests upon.

(D) parallel to the frictional force.

(E) parallel to the Earth's axis.

5. A horizontal beam of length 10 m and weight 200 N is attached to a wall as shown. The far end is supported by a cable which makes an angle of 60° with respect to the beam. A 500 N person stands 2 m from the wall. Determine the tension in the cable.

    (A) 0 N

    (B) 700 N

    (C) 500 N

    (D) 231 N

    (E) 808 N

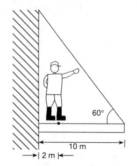

6. An object of mass M is suspended from three cables as shown below. The tension on cable B is

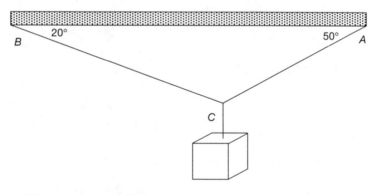

    (A) greater than Mg.

    (B) less than Mg.

    (C) equal to Mg.

    (D) equal to the tension on cable A.

    (E) There is insufficient information to obtain an answer.

7. Find the tension T_2 in cord 2 for the system drawn below. The system is in equilibrium.

    (A) 19.6 N

    (B) 39.2 N

    (C) 0 N

    (D) 17.0 N

    (E) 33.9 N

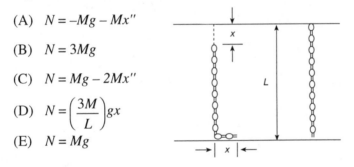

    $m = 2$ kg

8. A metallic chain of length L and M is vertically hung above a surface with one end in contact with it. The chain is then released to fall freely. If x is the distance covered by the end of the chain, how much force (exerted by the bottom surface) will the chain experience at any instance during the process?

    (A) $N = -Mg - Mx''$

    (B) $N = 3Mg$

    (C) $N = Mg - 2Mx''$

    (D) $N = \left(\dfrac{3M}{L}\right)gx$

    (E) $N = Mg$

9. Force resolution

    (A) decomposes a vector into its x, y, and z components.

    (B) is another term for frictional forces.

    (C) is another term for normal forces.

(D) decomposes a vector into its magnitude and direction.

(E) None of these.

10. A charged pith ball of mass 2 g is suspended on a massless string in an electric field

$$E = (3x + 4y) \times 10^5 \text{ N/C}.$$

If the ball is in equilibrium at $\theta = 57°$, then find the tension in the string.

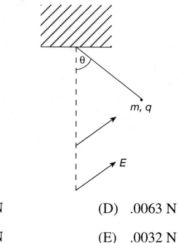

(A) .0500 N   (D) .0063 N

(B) .0250 N   (E) .0032 N

(C) .0125 N

ANSWER KEY

1.	(E)	6.	(B)
2.	(C)	7.	(E)
3.	(C)	8.	(D)
4.	(C)	9.	(A)
5.	(D)	10.	(C)

CHAPTER 14

Fluid Statics

14.1 Pressure in a Static Fluid

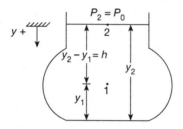

Figure 14.1 Pressure in a Static Fluid

$$p = p_0 + \rho g h \quad \rightarrow \text{units:} \frac{\text{newtons}}{\text{meters}^2} = \frac{\text{force}}{\text{area}}$$

p = pressure at point 1 at any level

p_0 = surface pressure (i.e., atmospheric pressure)

ρ = density

g = gravitational constant

h = distance from surface

Problem Solving Example:

Q A rectangular tank 6.0 by 8.0 m is filled with gasoline to a depth of 8.0 m. The pressure at the surface of the gasoline is 41,607 N/m². (The density of gasoline is 70.18 kg/m³.) Find the pressure at the bottom of the tank and the force exerted on the bottom.

A The total pressure at the tank's bottom is the sum of the air pressure at the surface of the fluid and the pressure due to the gasoline above the tank bottom.

$$P_{air} = 41,607 \text{ N/m}^2$$

To find the pressure on the bottom of the tank due to the gasoline, we note that the pressure is equal to the force on the bottom of the tank divided by the area of the bottom.

$$P_{gas} = \frac{F}{A}$$

But $F_{gas} = \rho g V$, where ρ is the density of the gasoline, g is 9.8 m/s², and V is the volume of the gasoline in the tank. Hence,

$$P_{gas} = \frac{\rho g V}{A}$$

But $V = hA$, where h is the height of the gasoline in the tank. Therefore,

$$P_{gas} = \rho g h = \left(70.18 \text{ kg/m}^3\right)\left(9.8 \text{ m/s}^2\right)\left(8 \text{ m}\right)$$
$$P_{gas} = 5,502 \text{ N/m}^2$$

Hence,

$$P_{total} = 5,502 + 41,607 \text{ N/m}^2$$
$$P_{total} = 47,109 \text{ N/m}^2$$

Noting that
$$P_{total} = \frac{F_{total}}{A}$$

$$F_{total} = P_{total}\, A$$
$$= (47,109 \text{ N/m}^2)\,(48 \text{ m}^2)$$
and
$$= 2,261,267 \text{ N}$$

14.2 Archimedes' Principle (Buoyancy)

Buoyancy Force, F_B

$$F_B = W_{df}$$

F_B = buoyancy force

W_{df} = weight of displaced fluid

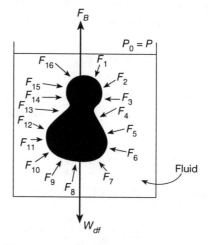

$$\overline{W}_{df} = \overline{F}_B = \overline{F}_1 + \overline{F}_2 + \overline{F}_3 + \ldots + \overline{F}_{16}$$

Figure 14.2 Buoyancy Force

Problem Solving Example:

What is the apparent loss of weight of a cube of steel 1 cm on a side submerged in water if the weight density of H_2O is 0.8022 kg/m^3?

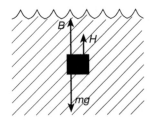

 The situation is as depicted in the diagram. Assume that a diver's hand exerts an upward force H on the steel, which is just large enough to keep the block in equilibrium. In this case, from Newton's second law, the net force on the block is zero and

$$B + H - mg = 0 \qquad (1)$$

where B is the buoyant force of the water on the steel. By Archimedes' Principle, B is equal to the weight of water displaced by the block. Hence,

$$B = \rho_w gV \qquad (2)$$

where V is the block's volume, and ρ_w is the density of water. Therefore, solving (1) for H, and using (2),

$$H = mg - B = mg - \rho_w gV$$

But m may be written as

$$m = \rho_s V \qquad (3)$$

where ρ_s is steel's density. Finally, then,

$$H = \rho_s gV - \rho_w gV$$

But H is the apparent weight of the steel, since this is the force we exert on the block to keep it in equilibrium. The weight of the block outside the water is given by (3) as

$$mg = \rho_s gV$$

The difference between "apparent" and "actual" weights is then

$$\Delta = H - mg = \left(\rho_s gV - \rho_w gV\right) - \rho_s gV$$
$$|\Delta| = \rho_w gV$$

which is B. Hence,

$$|\Delta| = (0.8022\,\text{kg/m}^3)(0.15\,\text{m})^3(9.8\,\text{m/s}^2)$$
$$|\Delta| = 0.027\,\text{kg}$$

14.3 Measurement of Pressures

a) Atmospheric Pressure, P_0 (Mercury)

$$P_0 = Pgh$$

P = density of mercury

g = acceleration due to gravity

h = height change of the mercury

b) Gauge Pressure

$$P - P_0 = \text{gauge pressure}$$

Problem Solving Example:

 Compute the atmospheric pressure on a day when the height of the barometer is 76.0 cm.

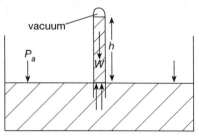

The height of the mercury column of the barometer depends on density ρ and g, as well as on the atmospheric pressure. Hence, both the density of mercury and the local acceleration of gravity must be known. The density varies with the temperature and g with the latitude and elevation above sea level. All accurate barometers are provided with a thermometer and with a table or chart from which corrections for temperature and elevation can be found. Let us assume $g = 980$ cm/sec² and $\rho = 13.6$ gm/cm³. The pressure due to the atmosphere supports the weight of mercury in the column of the barometer (see the figure on previous page). If the cross-sectional area of the column is A, then the weight of mercury in the column is

$$W = mg$$

where m is the mass of the mercury. Since

$$\text{Density}\,(\rho) = \frac{\text{mass}}{\text{volume}}$$

then
$$W = \rho(Ah)g$$

where $V = Ah$, h being the height of mercury in the column. Therefore,

$$\frac{W}{A} = \rho g h$$

is the pressure due to the weight of the mercury acting downward. It must equal P_a for equilibrium to be maintained in the fluid. (See figure on previous page.) Hence,

$$P_a = \rho g h = 13.6\,\frac{\text{gm}}{\text{cm}^3} \times 980\,\frac{\text{cm}}{\text{sec}^2} \times 76\,\text{cm}$$

$$= 1,013,000\,\frac{\text{dynes}}{\text{cm}^2}$$

(about a million dynes per square centimeter). In British engineering units,

$$76 \text{ cm} = 30 \text{ in} = 2.5 \text{ ft}$$

$$\rho g = 850 \frac{\text{lb}}{\text{ft}^3}$$

$$P_a = 2{,}120 \frac{\text{lb}}{\text{ft}^2} = 14.7 \frac{\text{lb}}{\text{in}^2}$$

CHAPTER 15

Fluid Dynamics

15.1 Equation of Continuity

$$\rho_1 A_1 v_1 = \rho_2 A_2 v_2 = \text{constant}$$

ρ = density

A = cross-sectional area

v = velocity

Steady Incompressible Flow $(\rho_1 = \rho_2)$

$$A_1 v_1 = A_2 v_2 = \text{constant}$$

Problem Solving Example:

Q Water flows at the rate of 300 ft³/min through an inclined pipe as shown in the figure on the next page. At A, where the diameter is 12 in., the pressure is 15 lb/in². What is the pressure at B, where the diameter is 6.0 in. and the center of the pipe is 2.0 ft. lower than at A?

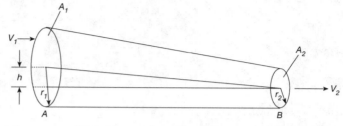

A The mass of liquid entering the tube at point A, in a time Δt, should be equal to the mass leaving it at point B in Δt. Let the velocities and the densities at A and B be v_1, v_2 and ρ_1, ρ_2, respectively. In a time Δt, the liquid entering at A moves a distance $v_1\Delta t$; hence, the volume of liquid entering is $Av_1\Delta t$. Therefore, the mass of this fluid is $\rho_1 v_1 A_1 \Delta t$. Similarly, the mass of liquid leaving at B is $\rho_2 v_2 A_2 \Delta t$. Dividing both terms by Δt, we get the continuity equation for the flow

$$\rho_1 v_1 A_1 = \rho_2 v_2 A_2$$

For all practical purposes, we can assume liquids to be incompressible; therefore, the density of liquid during flow remains constant

$$\rho_1 = \rho_2$$

and we have

$$v_1 A_1 = v_2 A_2.$$

For this problem

$$A_1 v_1 = \frac{300 \text{ ft}^3/\text{min}}{60 \text{ sec/min}} = 5 \text{ ft}^3/\text{sec}$$

Hence, $\quad v_1 = \dfrac{A_1 v_1}{A_1} = \dfrac{A_1 v_1}{\pi r_1^2} = \dfrac{5 \text{ ft}^3/\text{sec}}{3.14 \times (1/2 \text{ ft})^2} = 6.4 \text{ ft/sec}$

$$v_2 = \frac{A_1}{A_2} v_1 = \frac{\pi(1/2 \text{ ft})^2}{\pi(1/4 \text{ ft})^2} v_1 = 4 v_1$$

$$v_2 = 25.6 \text{ ft/sec}$$

The pressure P_1 at point A is

$$P_1 = \left(15 \text{ lb/in}^2\right)\left(144 \text{ in}^2/\text{ft}^2\right) = 2,200 \text{ lb/ft}^2$$

The weight density of water $D = 62.4$ lb/ft^3, and therefore

$$\rho = 1.94 \text{ slugs/ft}^3.$$

The Bernoulli equation for the pressure at A and B is

$$P_1 - P_2 = \rho g h + \frac{1}{2}\rho\left(v_2^2 - v_1^2\right)$$

where h is the difference between the heights of the centers of the cross-sections at the two ends of the pipe. The pressure at point B is

$$P_1 = P_2 + \rho g \left(h_1 - h_2\right) + \frac{\rho}{2}\left(v_1^2 - v_2^2\right)$$

$$= 2,200 \text{ lb/ft}^2 + \left(62.4 \text{ lb/ft}^3\right)(2.0 \text{ ft})$$

$$+ \frac{1.94 \text{ slug/ft}^3}{2}\left[(6.4)^2 - (25.6)^2\right]\text{ft}^2/\sec^2$$

$$= 2,200 \text{ lb/ft}^2 + 125 \text{ lb/ft}^2 - 596 \text{ lb/ft}^2$$

$$= 1,729 \text{ lb/ft}^2$$

$$= 12 \text{ lb/in}^2$$

15.2 Bernoulli's Equation

Steady Flow

$$p_1 + \frac{1}{2}\rho v_1^2 + \rho g y_1 = p_2 + \frac{1}{2}\rho v_2^2 + \rho g y_2 = \text{constant}$$

where $\quad p + \rho g y = $ static pressure

$$\frac{1}{2}\rho v^2 = \text{dynamic pressure}$$

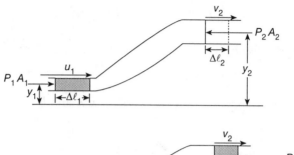

Figure 15.1 A Portion of Fluid (Shaded Area) Travels Through a Section of Pipe with a Changing Cross-sectional Area

Problem Solving Example:

Q Water flows into a water tank of large cross-sectional area at a rate of 10^{-4} m³/s, but flows out from a hole of area 1 cm², which has been punched through the base. How high does the water rise in the tank?

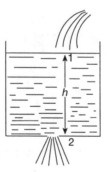

Bernoulli's Equation

A When water reaches its maximum height in the tank, the pressure head is great enough to produce an outflow exactly equal to the inflow. Equilibrium is then reached and the water level in the tank stays constant.

Since the cross-sectional area of the tank is large in comparison with the area of the hole, the water in the tank may be considered to have zero velocity. Further, the air above the tank and outside the holes are each at atmospheric pressure. Apply Bernoulli's theorem

$$p_1 + \rho g y_1 + \frac{1}{2}\rho v_1^2 = p_2 + \frac{1}{2}\rho g y_2 + \frac{1}{2}\rho v_2^2$$

with point 1 at the surface of the water at a height h above the hole and point 2 the hole itself. Then

$$p_a + \rho g h + 0 = p_a + 0 + \frac{1}{2}\rho v^2$$

where v is the velocity of efflux from the hole. Hence,

$$v = \sqrt{2gh}$$

But at equilibrium v is the rate of influx divided by the area of the hole. That is

$$v = \frac{10^{-4}\,\text{m}^3/\text{s}}{10^{-4}\,\text{m}^2} = 1\,\text{m/s}$$

Therefore, the maximum height of water in the tank is

$$h = \frac{v^2}{2g} = \frac{1^2\,\text{m}^2/\text{s}^2}{2 \times 9.8\,\text{m/s}^2} = 5.1\,\text{cm}$$

15.3 Viscosity

$$F = \eta \frac{Av}{L} \quad \rightarrow \text{units: kg/m} \bullet \text{s}^2$$

η = coefficient of velocity

F = the forces that tend to drag the liquid left and right

A = area of liquid over which forces are applied

L = transverse dimension

v = velocity

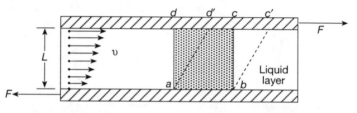

Figure 15.2 Laminar Flow of a Viscous Liquid

Problem Solving Example:

Q An old-fashioned water clock consists of a circular cylinder 10 cm in diameter and 25 cm high with a vertical capillary tube 40 cm in length and 0.5 mm in diameter attached to the bottom. The viscosity of water is 0.01 poise. What is the distance between hour divisions at the top of the vessel and at the bottom of the vessel?

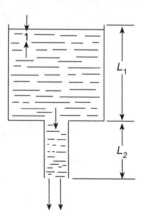

A The total volume of liquid Q, which flows across the entire cross-section of a cylindrical tube in time t, is given by Poiseuille's law

$$Q = \frac{\pi}{8} \frac{R^4}{\eta L} \Delta p t$$

where R is the radius, η is the viscosity of the liquid, and Δp is the pressure difference between the two cross-sectional surfaces separated by a distance L. The water in the capillary flows as a result of the pressures due to the water in the cylinder and its own weight, as shown in the figure. Under its own weight, it would have a rate of flow given by

$$\frac{Q_1}{t} = \frac{\pi}{8} \frac{R^4}{\eta L_2} (\rho g L_2)$$

where ρ is the density of water and g is the gravitational acceleration. The weight of the water exerts a pressure $\rho g L$, on the upper cross-section of the capillary and gives rise to another pressure difference between the two ends of the capillary. The rate of flow due to this pressure difference is

$$\frac{Q_2}{t} = \frac{\pi}{8} \frac{R^4}{\eta L_2} (\rho g L_1)$$

The total rate of flow is therefore

$$\frac{Q}{t} = \frac{Q_1}{t} + \frac{Q_2}{t} = \frac{\pi}{8} \frac{R^4}{\eta L_2} \rho g (L_1 + L_2)$$

The quantity of water, Q, flowing from the capillary in time t causes a drop in the level of the cylinder, h. The area of the cylinder is A, and thus $Q/t = Ah/t$.

$$\therefore \frac{h}{t} = \frac{\pi}{8A} \frac{R^4}{\eta} \frac{\rho g (L_1 + L_2)}{L_2}$$

When the cylinder is full, $L_1 = 25$ cm, and

$$\frac{h}{t} = \frac{\pi \times \left(0.25 \times 10^{-3}\,\text{m}\right)^4 \times 10^3\,\text{kg/m}^3 \times 9.8\,\text{m/s}^2 \times 0.65\,\text{m}}{8 \times \pi / 4 \times 10^{-2}\,\text{m}^2 \times 10^{-3}\,\text{Ns}/\text{m}^2 \times 0.40\,\text{m}}$$

$$= 3.11 \times 10^{-6}\,\text{m/s} = 1.12\,\text{cm/hr}$$

When the cylinder is empty, $L_1 = 0$ cm, and

$$\frac{h'}{t} = \frac{\pi R^4}{8 A \eta L_2}\,\rho g L_2 = \frac{\pi R^4}{8 A \eta L_2}\left(L_1 + L_2\right) \cdot \frac{L_2}{L_1 + L_2}$$

$$= \frac{h}{t}\,\frac{L_2}{L_1 + L_2} = \frac{h}{t}\,\frac{0.40}{0.65}$$

$$= 1.12 \times \text{cm/hr.} \cdot \frac{0.40}{0.65} = 0.69\,\text{cm/hr.}$$

Thus, hour divisions are separated by 1.12 cm at the top and 0.69 cm at the bottom. Note that L_1 varies slightly during the hour and, to be quite exact, an integration ought to be performed. The error involved is, however, slight, since the variation in L_1 is very small in comparison with $L_1 + L_2$.

15.4 Reynolds Number: N_R

$$N_R = \frac{\rho v D}{\eta}$$

ρ = density of the fluid

v = average velocity

D = diameter of the pipe

η = coefficient of viscosity

Problem Solving Example:

 Spherical particles of pollen are shaken up in water and allowed to stand. The depth of water is 2 cm. What is the diameter of the largest particles remaining in suspension 1 hour later? Density of pollen = 1.8 g/cm³.

 The terminal velocity of the particles after they are allowed to settle will very quickly be reached. After 1 hour the only particles left in suspension are those that take longer than 1 hour to fall 2 cm. The larger, heavier particles have already settled. The particles that have just not settled are those that take exactly 1 hour to fall 2 cm. That is

$$v_T = \frac{2\ \text{cm}}{(1\ \text{hr})(3,600\ \text{s/hr})} = \frac{1}{1,800}\ \text{cm/sec}$$

We need another expression for the terminal velocity. Stoke's law states that when a sphere moves through a viscous fluid at rest, the resisting force f exerted by the fluid on the sphere is given

$$f = 6\pi\eta rv$$

where η is the viscosity of the fluid, r is the radius of the sphere, and v is its velocity with respect to the fluid. The other forces that act on the sphere are its weight mg and the upward buoyant force B of the fluid. Let ρ be the density of the sphere and ρ' the density of the fluid. Then

$$mg = \frac{4}{3}\pi r^3 \rho g$$

$$B = \frac{4}{3}\pi r^3 \rho' g \qquad \text{(Archimedes' principle)}$$

The net force on the sphere equals the product of its mass and acceleration. Taking the downward direction as positive

$$mg - B - R = ma$$

$$a = g - \frac{B+R}{m}$$

Assuming the initial velocity is zero, this net acceleration imparts a downward velocity to the sphere. As this velocity increases, so does the retarding force. At some terminal v_T, the retarding force has increased an amount such that the downward acceleration equals zero. At this point, the velocity of the sphere stays constant and is found by setting the acceleration equal to zero. Then

$$mg = B + R$$

$$\frac{4}{3}\pi r^3 \rho g = \frac{4}{3}\pi r^3 \rho' g + 6\pi\eta r v_T$$

$$v_T = \frac{2}{9}\frac{r^2 g}{\eta}(\rho - \rho')$$

The radius of the largest particles still just in suspension is thus given by

$$r^2 = \frac{9}{2}\frac{\eta v_T}{g(\rho - \rho')}$$

$$= \frac{9}{2}\frac{1 \times 10^{-2}\text{ poise} \times \frac{1}{1,800}\text{ cm}}{980\text{ cm/s}^2(1.8 - 1)\text{ g/cm}^3} = \frac{10^{-4}}{64 \times 49}\text{ cm}^2$$

$$d = 2r = \frac{2 \times 10^{-2}}{8 \times 7}\text{ cm} = 3.57 \times 10^{-4}\text{ cm}$$

CHAPTER 16

Temperature

16.1 The Kelvin, Celsius, and Fahrenheit Scales

a) Absolute (Kelvin) Temperature

$T_{tr} = 273.16$ K

T_{tr} = temperature at which H_2O (water) can coexist as a vapor, a liquid, and a solid (triple point)

K = unit of temperature interval = one "Kelvin" (K)

b) Celsius Temperature

$$t_c = \left(t_k - 273.15\right)^\circ C$$

t_c = Celsius temperature

t_k = Kelvin temperature

c) Fahrenheit Temperature

$$t_F = \left(\frac{9}{5}t_c + 32\right)^\circ F$$

t_F = Fahrenheit temperature, in "degrees Fahrenheit" (°F)

t_c = Celsius temperature

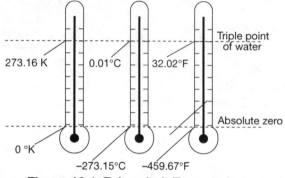

Figure 16.1 Fahrenheit Temperature

Problem Solving Example:

 A Celsius thermometer indicates a temperature of 36.6°C. What would a Fahrenheit thermometer read at that temperature?

The relationship between the Celsius and Fahrenheit scales can be derived from a knowledge of their corresponding values at the freezing and boiling points of water. These are 0°C and 32°F for freezing and 100°C and 212°F for boiling. The temperature change between the two points is equivalent for the two scales and a temperature difference of 100 Celsius degrees equals 180 Fahrenheit degrees. Therefore, one Celsius degree is $^9/_5$ as large as one Fahrenheit degree. We can then say

$$°F = \frac{9}{5}°C + B$$

where B is a constant. To find it, substitute the values known for the freezing point of water

$$32° = \frac{9}{5} \times 0° + B = B$$

We therefore have

$$°F = \frac{9}{5}°C + 32°$$

For a Celsius temperature of 36.6°, the Fahrenheit temperature is

$$F = \frac{9}{5} \times 36.6° + 32°$$
$$= 65.9° + 32° = 97.9°$$

16.2 Coefficients of Thermal Expansion

a) Coefficient of Linear Expansion

$$\alpha = \frac{\Delta L}{L_0} \frac{1}{\Delta t}$$

$L = L_0 (1 + \alpha \Delta t)$

b) Surface Expansion

$$A = A_0(1 + \alpha \Delta t) \quad \rightarrow \text{units: meters}^2$$
$$\gamma = 2\alpha$$

γ = coefficient of surface expansion

A = area of temperature t

A_0 = area at t_0

Δt = change in temperature

c) Volume Expansion

$$V = V_0(1 + \beta \Delta t) \quad \rightarrow \text{units: meters}^3$$
$$\beta = 3\alpha$$

β = coefficient of volume expansion

V = volume at temperature t

V_0= volume at t_0

Δt= change in temperature

Problem Solving Example:

 An iron steam pipe is 70 m long at 0°C. What will its increase in length be when heated to 100°C? ($\alpha = 10 \times 10^{-6}$ per Celsius degree.)

 The change in length, ΔL, of a substance due to a temperature change is proportional to the change, ΔT, and to the original length, L_0, of the object

$$\Delta L = \alpha L_0 \, \Delta T$$

where α is the proportionality constant and is called the coefficient of linear expansion.

$L_0 = 200$ ft

$\alpha = 10 \times 10^{-6}$ per C°

$T = 100°C$

$T_0 = 0°C$

Increase in length $= \Delta L = \alpha L_0 \Delta T$
$$= \left(10 \times 10^{-6}\right)(200)(100)$$
$$= 0.20 \text{ ft}$$

Quiz: Fluid Statics—Temperature

1. Absolute temperature is best described as a measure of the

   (A) speed of molecules.

   (B) mass of molecules.

   (C) pressure between molecules.

   (D) number of molecules.

   (E) average translational kinetic energy of molecules.

2. The internal energy of water is determined by which of the following?

   I. Temperature

   II. Phase

   III. Mass

   (A) I only.   (D) I and II only.

   (B) II only.   (E) I, II, and III.

   (C) III only.

3. The relationship between the pressure and volume of an ideal gas when temperature is held constant is

   (A) logarithmic.   (D) direct.

   (B) geometric.   (E) ratio.

   (C) inverse.

4. The amount of energy that must be added to a unit quantity of a material to raise its temperature by one degree is called the

(A) specific heat.

(D) potential energy.

(B) internal energy.

(E) thermal energy.

(C) heat content.

5. Thermal energy added to a sample of a substance that changes the phase of the substance is called

(A) heat of formation.

(D) heat of reaction.

(B) calorimetry.

(E) heat content.

(C) latent heat.

6. A typical liquid-filled thermometer works due to the fact that as the thermometer heats up, its

(A) liquid molecules grow larger.

(B) liquid molecules require more space between themselves.

(C) liquid molecules multiply.

(D) outer tube contracts and squeezes liquid into the thermometer.

(E) Both (B) and (D).

7. A large square sheet of copper with a small hole in it is heated from 30°C to 130°C. Which of the following will occur?

(A) The area of the sheet increases and the diameter of the hole increases.

(B) The area of the sheet increases and the diameter of the hole decreases.

(C) The area of the sheet decreases and the diameter of the hole increases.

(D) The area of the sheet decreases and the diameter of the hole decreases.

(E) The area of the sheet increases and the diameter of the hole remains constant.

8. The number of calories needed to melt 1 gram of ice at 0°C is the same as the amount needed to

(A) vaporize 1 gm of water at 100°C.

(B) condense 1 gm of steam at 100°C.

(C) raise the temperature of 1 gm of water 80°C.

(D) melt 1 gm of copper at 0°C.

(E) melt 5 gm of ice at 0°C.

9. A wide range of temperatures is currently accessible in the laboratory and through observation. Which one of the following is NOT a true statement about temperature?

(A) 0 K is the absolute zero of temperature.

(B) 20 K is the vaporization point of hydrogen.

(C) 144 K is the vaporization point of nitrogen.

(D) 1,234 K is the fusion temperature of silver.

(E) 6,000 K is the sun's surface temperature.

10. Which one of the following comments regarding the addition of heat to a system is false?

(A) The conduction of heat always involves a transfer of heat.

(B) The addition of heat always causes a rise in temperature.

(C) In a perfectly isolated system, any heat lost by one part of the system always equals the heat gained by another part of the same system.

(D) Heat and temperature are not the same thing.

(E) Temperature can change even though no heat is lost or gained.

ANSWER KEY

1.	(E)		6.	(B)
2.	(E)		7.	(A)
3.	(C)		8.	(C)
4.	(A)		9.	(C)
5.	(C)		10.	(B)

CHAPTER 17

Heat and the First Law of

Thermodynamics

17.1 Quantity of Heat and Specific Heat

a) Heat Capacity

$$C = \frac{Q}{\Delta T} \qquad \rightarrow \text{units: cal/gm} \cdot {}^{\circ}\text{C}$$

C = heat capacity

Q = heat applied

ΔT = change in temperature

b) Specific Heat

$$c = \frac{Q}{m\Delta T} \qquad \rightarrow \text{units: cal/gm} \cdot {}^{\circ}\text{C}$$

c = specific heat

Q = heat applied

m = mass of heated substance

ΔT = change in temperature

Problem Solving Example:

 How many calories of heat are required to raise 1,000 grams of water from 10°C to 100°C?

 The temperature rise is 90°C. The number of calories needed is

$$1,000 \times 90° = 90,000 \text{ calories}$$

since one calorie is required to raise the temperature of one gram of water one degree centigrade.

17.2 Heat Conduction

$$H = -kA\frac{dT}{dx}$$

H = heat flow rate

k = thermal conductivity

A = cross-sectional area

dT = temperature difference

dx = thickness

$\dfrac{dT}{dx}$ = thermal gradient

For a rod of constant cross-sectional area

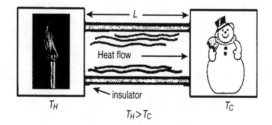

Figure 17.1   Heat Conduction

$$H = -kA\frac{T_H - T_c}{L}$$

Problem Solving Example:

Q On either side of a pane of window glass, temperatures are 70°F and 0°F. How fast is heat conducted through such a pane of area 2,500 cm² if the thickness is 2 mm?

A The equation of heat conduction is

$$\frac{dQ}{dt} = -KA\frac{dT}{dx} \qquad (1)$$

where dQ/dt is the rate at which heat is transferred across a cross-section A of a material with coefficient of thermal conductivity K. dT/dx is the temperature gradient in the material.

In the steady state, the temperature at each point of the material remains constant in time. Hence, the rate of heat transfer across a cross-section is the same at all cross-sections. As a result of (1), dT/dx must be the same at all cross-sections. If T_1 is the temperature at a cross-section at x_1, and T_2 is the temperature at x_2, we obtain

$$\frac{dT}{dx} = \frac{\Delta T}{\Delta x} = \frac{T_2 - T_1}{x_2 - x_1} \qquad (2)$$

(Note that this is a direct consequence of the fact that dT/dx is constant.) Using (2) in (1),

$$\frac{dQ}{dt} = -KA\left(\frac{T_2 - T_1}{x_2 - x_1}\right) = KA\left(\frac{T_1 - T_2}{x_2 - x_1}\right)$$

But $x_2 - x_1$ is equal to L, the length of the material across which the heat conduction is taking place.

$$\frac{dQ}{dt} = \frac{KA(T_1 - T_2)}{L}$$

For the pane of glass,

$$T_1 = 70°F \qquad @\ x_1 = 0 \text{ mm}$$

$$T_2 = 0°F \qquad @\ x_2 = 2 \text{ mm}$$

Furthermore, $K = .0015$ cal/cm $\times$ s $\times$ °C for glass, whence

$$\frac{dQ}{dt} = \frac{(.0015 \text{ cal/cm} \bullet \text{s} \bullet °\text{C})(2,500 \text{ cm}^2)(70°\text{F} - 0°\text{F})}{(2 \text{ mm} - 0 \text{ mm})}$$

Since $70°F = 5/9\ 70°C = 350°C/9$, we obtain

$$\frac{dQ}{dt} = \frac{(.0015 \text{ cal/cm} \bullet \text{s} \bullet °\text{C})(2,500 \text{ cm}^2)(350°\text{C}/9)}{(.2 \text{ cm})}$$

$$\frac{dQ}{dt} = 729 \text{ cal}/\text{s}$$

Note that, by convention, temperature decreases as x increases. Hence, $T_2 < T_1$ and $x_2 > x_1$. As a result, $dQ/dt > 0$ in the direction in which $dT/dx < 0$.

17.3 Thermal Resistance, *R*

$$R = \frac{L}{K}$$

Heat Conduction

$$H = A\frac{T_h - T_c}{R}$$

L = length of rod

H = heat flow rate

A = cross-sectional area

K = Thermal Conductivity

R = thermal resistance

T_h = higher temperature

T_c = *lower temperature*

Problem Solving Example:

A copper kettle, the circular bottom of which is 0.1524 m = 15.2 cm in diameter and 0.1575 cm thick, is placed over a gas flame. On assuming that the average temperature of the outer surface of the copper is 101°C and that the water in the kettle is at its normal boiling point, how much heat is conducted through the bottom in 5.0 sec? The thermal conductivity may be taken as 6.249 kg cal/cm.

The heat Q conducted through the bottom of the kettle in time t is given by

$$Q = K \, At \frac{\Delta T}{\Delta x}$$

where $\Delta T/\Delta x$ is the temperature gradient in C°/cm, K is the thermal conductivity, and A is the area of the bottom in cm². We have

$$A = \pi r^2 = \pi \left(\frac{3.0}{12} \text{ft} \right)^2 = 0.20 \text{ ft}^2$$

$$t = 5.0 \text{ sec} = \frac{5.0}{3,600} \text{ hr} = 0.0014 \text{ hr}$$

The temperature on the inside of the bottom of the kettle is the same as that of boiling water (212°F). Since the temperature on the outside of the bottom is 214°F and the thickness of the bottom of the kettle is 0.062 in, the temperature gradient across the bottom is

$$\frac{\Delta T}{\Delta x} = \frac{214°F - 212°F}{0.062 \text{ in}} = 32°F/in$$

The heat conducted through the bottom is then

$$Q = \left(2,480 \frac{\text{Btu}}{\text{ft}^2 \text{ hr °F/in}} \right) \left(0.20 \text{ ft}^2 \right) \left(0.0014 \text{ hr} \right) \left(32°\text{F/in} \right)$$

$$= 22 \text{ Btu}$$

17.4 Convection

$$H = LA\Delta T \quad \rightarrow \text{units:} \left(\text{cal/sec} \right) \bullet \text{cm}^2 \bullet \text{deg}$$

L = convection coefficient

H = heat convection current

A = surface area

ΔT = change in temperature

17.5 Thermal Radiation

$$*R = e\sigma T^4$$

R = energy radiated per unit area

σ = Stefan's constant, 5.6703×10^{-8} w/m²K⁴

T = kelvin temperature

e = emissivity of the surface

*Also known as Stefan's law.

Problem Solving Example:

Q The solar constant, or the quantity of radiation received by the earth from the sun, is 0.14 cm^{-2}. Assuming that the sun may be regarded as an ideal radiator, calculate the surface temperature of the sun. The ratio of the radius of the earth's orbit to the radius of the sun is 216.

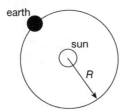

 To calculate the temperature of the sun, T, we use Stefan's law

$$R = e\sigma T^4$$

Here, e is the emissivity of the radiator, σ is a constant, T is the temperature of the radiator in Kelvin, and R is the rate of emission of radiant energy per unit area of the radiator. Hence,

$$T^4 = \frac{R}{e\sigma} \tag{1}$$

Regarding the sun as an ideal radiator, $e = 1$. Furthermore,

$$R = \frac{P}{A}$$

where A is the surface area of the sun, and P is the power provided by the sun as a result of radiation. Using these facts in (1),

$$T^4 = \frac{P}{\sigma A} \tag{2}$$

Now, the power per unit area intercepted by the earth is

$$\frac{P}{A'} = .14 \ \text{W/cm}^2$$

where A' is the surface area of a sphere having a radius equal to that of the earth's orbit. Hence,

$$P = (.14 \ \text{W/cm}^2) \ A' \tag{3}$$

Using (3) in (2),

$$T^4 = \frac{(.14 \text{ W/cm}^2) A'}{\sigma A}$$

Now

$$\frac{A'}{A} = \frac{4\pi r^2}{4\pi R^2}$$

where r and R are the radius of the earth's orbit and the radius of the sun, respectively. Then,

$$T^4 = \frac{(.14 \text{ W/cm}^2)}{\sigma} \left(\frac{r}{R}\right)^2$$

or

$$T^4 = \frac{0.14 \text{ W} \cdot \text{cm}^{-2}}{5.6 \times 10^{-12} \text{ W} \cdot \text{cm}^{-2} \cdot \text{K deg}^{-4}} \times (216)^2$$

$$\therefore T = 5.84 \times 10^3 \text{ K}.$$

17.6 Heat and Work

$$W = \int_{V_i}^{V_f} p \, dV$$

W = work done

V_i = initial volume

V_f = final volume

p = pressure

dV = differential change in volume

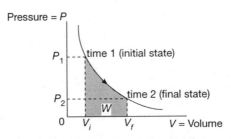

Figure 17.2 Heat and Work

Problem Solving Example:

One liter of an ideal gas under a pressure of 1 atm is expanded isothermally until its volume is doubled. It is then compressed to its original volume at constant pressure and further compressed isothermally to its original pressure. Plot the process on a p-V diagram and calculate the total work done on the gas. If 50 J of heat were removed during the constant-pressure process, what would be the total change in internal energy?

During an isothermal change, T is constant.

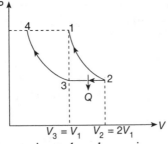

The work done on the gas in such a change is

$$W = -\int_{V_i}^{V_f} p\,dV$$

This is negative because the pressure on V_i the gas is in opposition to the volume change. From the ideal gas law, $pV = nRT$, we get

$$W = -\int_{V_i}^{V_f} \frac{nRT}{V} dV = -nRT \int_{V_i}^{V_f} \frac{dV}{V} = -nRT \ln \frac{V_f}{V_i}$$

$$= -p_i V_i \ln \frac{V_f}{V_i} = -p_i V_i \ln \frac{p_i}{p_f}$$

where the subscripts i and f refer to initial and final states, respectively.

Thus, the work done on the gas in the first change is (from (1) to (2), as shown in the figure on the previous page)

$$W_1 = -p_1 V_1 - p_i V_i \ln \frac{V_2}{V_1}$$

$$= -(1 \text{ atm} \times 1.013 \times 10^6 \text{ dynes/cm}^2 \text{ atm})$$

$$\times (1 \text{ liter} \times 10^3 \text{ cm}^3/\text{lit}) \times 1n\ 2$$

$$= -1.013 \times 10^6 \text{ dynes/cm}^2 \times 10^3 \text{ cm}^3 \times 1n\ 2$$

$$= -7.022 \times 10^8 \text{ ergs} -70.22 \text{ J}$$

Further, since the volume is doubled, by the application of Boyle's law $p_1 V_1 = p_2 V_2$, we see that the pressure is halved at (2)

$$p_2 = p_1 \frac{V_1}{V_2} = \frac{1}{2} p_1$$

The work done on the gas in the second change is (from (2) to (3)).

$$W_2 = -p_2 \int_{V_2}^{V_3} dV = -p_2 (V_3 - V_2) = -\frac{1}{2} p_1 (V_1 - 2V_1)$$

$$= \frac{1}{2} p_1 V_1$$

$$= \frac{1.013 \times 10^6 \text{ dynes/cm}^2 \times 10^3 \text{ cm}^3}{2 \times 10^7 \text{ ergs} / \text{J}} = 50.65 \text{ J}$$

The work done on the gas in the final change from ((3) to (4)) is

$$W_3 = -p_3 V_3 \ln \frac{p_3}{p_4} = -\frac{1}{2} p_1 V_1 \ln \frac{p_2}{p_1}$$

$$= \frac{1}{2} p_1 V_1 \ln \frac{1}{2} = \frac{1}{2} p_1 V_1 \ln 2$$

$$= \frac{1.013 \times 10^6 \text{ dynes /cm}^2 \times 10^3 \text{cm}^3 \times \ln(2)}{2 \times 10^7 \text{ ergs/J}} = 35.11 \text{ J}$$

The total work done on the gas is thus,

$$W_1 + W_2 + W_3 = (50.65 + 35.11 - 70.22) \text{ J} = 15.54) \text{ J}.$$

In the first and third processes, the temperature does not change. In an ideal gas the internal energy depends only on the temperature, so that no change of internal energy takes place in the first and third processes. Any work done on the gas in these changes is equal to the heat transfer taking place.

The second process is isobaric. The change in internal energy during the process is given by the first law of thermodynamics as $\Delta U = Q - W$, where Q is the heat energy added to the system and W the work done by the system. Hence,

$$\Delta U = -50 \text{ J} - (-50.65 \text{ J}) = +0.65 \text{ J}.$$

The internal energy thus increases by 0.65 J during the three processes.

17.7 The First Law of Thermodynamics

$$Q = u_f - u_1 + W$$

Q = heat absorbed

$u_f - u_1$ = change in internal energy

W = work done

Problem Solving Example:

 Three moles of a diatomic perfect gas are allowed to expand at constant pressure. The initial volume is 1.3 m³ and the initial temperature is 350K. If 10,000 Joules are transferred to the gas as heat, what are the final volume and temperature?

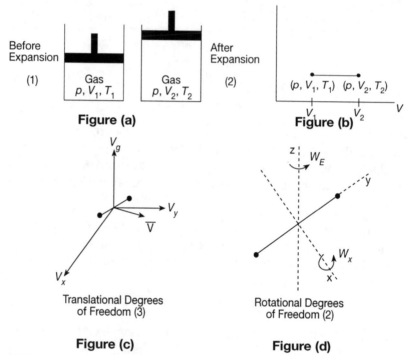

Before Expansion

(1)   Gas p, V_1, T_1 Gas p, V_2, T_2 (2)

After Expansion

(p, V_1, T_1) (p, V_2, T_2)

Figure (a)   **Figure (b)**

Translational Degrees of Freedom (3)

Rotational Degrees of Freedom (2)

Figure (c)   **Figure (d)**

 Since the process of heat addition (Q) occurs at constant pressure, we may write

$$Q = \mu C_p (T_2 - T_1)$$

where μ is the number of moles of gas in the system, C_p is the molar specific heat at constant pressure, and $T_2 - T_1$ is the temperature difference between the two equilibrium states (see Figures (a) and (b)). Now, Q is given (10,000 Joules), as is μ and T. Hence, we may solve for T_2 as a function of C_p. If we can calculate the value of C_p from kinetic theory, we will have obtained T_2. We now perform the appropriate calculation.

Consider a gas moving from one equilibrium state to another, via some thermodynamic process. We assume that the process occurs at constant volume. Using the First Law of Thermodynamics, we obtain

$$\Delta U = Q - W \tag{1}$$

Here, ΔU is the change in internal energy of the gas during the process, and Q and W are the heat added to and the work done by the gas, respectively, during the process. Writing (1) in differential form

$$dU = dQ - dW \tag{2}$$

In general, the element of work done by the gas in an expansion is, by definition,

$$dW = \overline{F} \cdot d\overline{s}$$

where $\overline{F}$ is the force the gas exerts on the piston (see Figure (a) on previous page) and $d\overline{s}$ is the element of distance the piston moves during the expansion. Since $\overline{F}$ acts perpendicular to the face of the piston (that is, $\overline{F}$ and $d\overline{s}$ are parallel), we obtain

$$dW = Fds \tag{3}$$

But F may be written in terms of the pressure p, that the gas exerts on the piston face of area A

$$F = pA$$

Using this in (3)

$$dW = pdsA = pdV$$

where dV is the differential change in volume of the gas during the expansion. Substituting this in (2), we find

$$dU = dQ - pdV \tag{4}$$

Applying this equation to the above-mentioned isovolumic process, we obtain

$$dU = dQ$$

since $dV = 0$. By definition, however,

$$dQ = \mu \, C_V \, dt$$

where C_V is the molar specific heat of the gas at constant volume, and dT is the differential temperature change the gas experiences due to the addition of heat dQ. Then

$$dU = \mu \, C_V \, dT \tag{5}$$

We now assume that the change in internal energy of a gas is only a function of the temperature difference experienced by the gas. Then, no matter what thermodynamic process the gas experiences, (5) holds.

Consider next an isobaric thermodynamic process. Again, we apply (4)

$$dU = dQ - p \, dV$$

Since the process occurs at constant pressure,

$$dQ = \mu \, CpdT \tag{6}$$

where Cp is the molar heat capacity at constant pressure. Furthermore, from the ideal gas law

$$pV = \mu RT$$

If p is constant,

$$p \frac{dV}{dT} = \mu R$$

or $$pdV = \mu RdT \tag{7}$$

Using (6), (7), and (5) in (4),

$$\mu C_v dT = \mu C_p dT - \mu RdT$$

or $$C_p = C_v + R \tag{8}$$

Equation (8) relates the molar specific heat at constant pressure to the molar specific heat at constant volume.

All the derivations up to now have been necessary in order to obtain certain relations involving molar specific heats, namely equations (5) and (8). We now turn to an examination of the internal energy of a diatomic gas. Each method of energy storage of a diatomic molecule is called a degree of freedom. If we view a diatomic molecule as being dumbbell-shaped, then it has 5 degrees of freedom (see Figures (c) and (d) on page 292). The molecule may move transitionally in three directions (x, y, z) with three kinetic energies

$$\left(\frac{1}{2} mv_x^2, \frac{1}{2} mv_y^2, \frac{1}{2} mv_z^2 \right).$$

Furthermore, it may rotate about three axes (x, y, z), again, with three kinetic energies

$$\left(\frac{1}{2} I_x\omega_x^2, \frac{1}{2} I_y\omega_y^2, \frac{1}{2} I_z\omega_z^2 \right).$$

However, the rotational kinetic energy about the y axis is negligible (see Figure (d) on page 292) because $I_y << I_x, I_z$. Hence, a diatomic molecule has five independent methods of energy absorption, or 5 degrees of freedom. Notice one important fact: each of these kinetic energy terms has the same form, mathematically. That is, they are all of the form of a positive constant times the square of a variable which has a domain extending from $-\infty$ to $+\infty$. The theorem of equipartition of energy tells us that, when Newtonian mechanics holds, and the number of gas particles is large, each term of this form has the same average value per molecule, namely, $\frac{1}{2} kT$. In other words, each degree of freedom of a gas molecule contributes an amount $\frac{1}{2} kT$ to the internal energy of the gas. For a diatomic gas, then the internal energy per molecule is

$$U_i = 5 \left(\frac{1}{2} kT \right) = \frac{5}{2} kT$$

The internal energy for μ moles of molecules is

$$U = \mu\, N_0 U_i = \frac{5}{2}\mu k N_0 T = \frac{5}{2}\mu R T$$

Using this in (5), we may solve for C_v

$$\frac{1}{\mu}\frac{d\left(\frac{5}{2}\mu R T\right)}{dT} = C_v$$

$$C_v = \frac{5}{2}R$$

Using this fact in (8),

$$C_p = C_v + R = \frac{5R}{2} + R = \frac{7}{2}R$$

Getting back to the original problem, use this value of C_p in the first equation

$$Q = \mu\, C_p (T_2 - T_1) = \frac{7}{2}\mu R (T_2 - T_1)$$

or $\qquad T_2 = T_1 + \dfrac{Q}{\dfrac{7}{2}\mu R} = 350\text{K} + \dfrac{10{,}000 \text{ Joules}}{\left(\dfrac{7}{2}\right)(3 \text{ moles})\left(8.31\,\dfrac{\text{Joules}}{\text{moleK}}\right)}$

$$T_2 = 350\text{K} + 114\text{K} = 464\text{K}$$

Using the ideal gas law for the two equilibrium states,

$$pV_1 = \mu R T_1$$
$$pV_2 = \mu R T_2$$

or $\qquad \dfrac{V_2}{V_1} = \dfrac{T_2}{T_1}$

whence $\qquad V_2 = \dfrac{T_2}{T_1}V_1 = \left(\dfrac{464}{350}\right)1.3\text{ m}^3 = 1.72\text{ m}^3$

17.8 Applications of the First Law of Thermodynamics

a) Adiabatic Process

$$U_f - U_i = -W$$

U_f = final internal energy

U_i = initial internal energy

W = work done

b) Isobaric (Constant Pressure) Process

$$W = p\left(V_f - V_i\right)$$

p = pressure

V_f = final volume

V_i = initial volume

c) Isochoric Process (Volume Unchanged)

$$Q = U_f - U_i$$

Q = heat absorbed

Problem Solving Example:

Q A cylinder contains an ideal gas at a pressure of 2 atm, the volume being 5 liters at a temperature of 250K. The gas is heated at constant volume to a pressure of 4 atm, and then at constant pressure to a temperature of 650K. Calculate the total heat input during these processes. For the gas, C_v is 21.0 J•mole^{-1}K deg^{-1}.

The gas is then cooled at constant volume to its original pressure and then at constant pressure to its original volume. Find the total heat output during these processes and the total work done by the gas in the whole cyclic process.

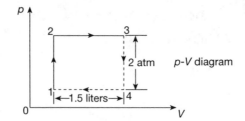

p-V diagram

The number of moles, n, originally present at point (1) in the *p-V* diagram can be calculated from the gas equation

$$n = \frac{pV}{RT}$$

$$= \frac{2 \text{ atm} \times 5 \text{ liters}}{0.0821 \text{ liter} \bullet \text{atm} \bullet \text{mole}^{-1} \bullet \text{K deg}^{-1} \times 250 \text{K}} = 0.487 \text{ mole}$$

The specific heat C_p at constant pressure is

$$C_p = C_v + R = (21.0 + 8.317) \text{ J} \bullet \text{mole}^{-1} \bullet \text{K deg}^{-1}$$

$$= 29.317 \text{ J} \bullet \text{mole}^{-1} \bullet \text{K deg}^{-1}$$

In going from (1) to (2), V is constant. Therefore, in the first change, P/T remains constant (from the universal gas law):

$$\frac{P_1}{T_1} = \frac{P_2}{T_2}$$

or

$$T_2 = T_1 \frac{P_2}{P_1}$$

Since

$$\frac{P_2}{P_1} = \frac{4 \text{ atm}}{2 \text{ atm}} = 2$$

we have

$$T_2 = 2T_1 = 2 \times 250 \text{K} = 500 \text{K}$$

The heat input along the change of state from (1) to (2) is

$$H_{12} = nC_v(T_2 - T_1)$$
$$= 0.487 \text{ mole} \times 21.0 \text{ J} \bullet \text{mole}^{-1} \bullet \text{K deg}^{-1} \times (500 - 250)\text{K}$$
$$= 2,558 \text{ J}$$

In the second change, from (2) to (3), P and therefore, V/T are constant

$$\frac{V_2}{T_2} = \frac{V_3}{T_3}$$

or

$$V_3 = \frac{T_3}{T_2}V_2 = \frac{T_3}{T_2}V_1$$
$$= \frac{650\text{K}}{500\text{K}} 5 \text{ lit} = 6.5 \text{ lit}$$

Heat input during this change is

$$H_{23} = nC_p(T_3 - T_2)$$
$$= 0.487 \text{ mole} \times 29.317 \text{ J} \bullet \text{mole}^{-1} \bullet \text{K deg}^{-1} \times (650 - 500)\text{K}$$
$$= 2,143 \text{ J}$$

The total heat input during these two processes is thus

$$H = H_{12} + H_{23} = 4,701 \text{ J}$$

During the change from (3) to (4), the gas cooled at constant volume. Hence,

$$\frac{P_3}{T_3} = \frac{P_4}{T_4}$$

or

$$T_4 = \frac{P_4}{P_3}T_3$$

Since
$$\frac{P_4}{P_3} = \frac{1}{2}$$

we get
$$T_4 = \frac{1}{2}T_3 = \frac{1}{2}650\text{K} = 325\text{K}$$

The heat rejected by the gas during this process is

$$H_{34} = nC_v(T_3 - T_4)$$
$$= 0.487 \text{ mole} \times 21.0 \text{ J} \bullet \text{mole}^{-1} \bullet \text{K deg}^{-1} \times (650 - 500)\text{K}$$
$$= 3,325 \text{ J}$$

In the second cooling process, from (4) to (1), P is kept constant:

$$\frac{V_4}{T_4} = \frac{V_1}{T_1}$$

or
$$T_1 = \frac{V_1}{V_4}T_4$$

Since
$$\frac{V_1}{V_4} = \frac{V_1}{V_3} = \frac{5 \text{ lit}}{6.5 \text{ lit}}$$

we get
$$T_1 = \frac{5}{6.5} \times 325\text{K} = 250\text{K}$$

as expected. The heat output during this change is

$$H' = H_{34} + H_{41} = 4,397 \text{ J}$$

The difference between heat input and heat output is 304 J. This must appear as work done by the gas, since the internal energy of the gas must be the same at the beginning and at the end of a cyclic process.

The mechanical work done during the cycle is given by

$$W = \int_1^2 pdV + \int_2^3 pdV + \int_3^4 pdV + \int_4^1 pdV,$$

which is the area enclosed by the rectangular figure in the p-V diagram.

This is a rectangle of height 2 atm and length 1.5 liters. The area under the curve is thus

$$W = 2 \times 1.013 \times 10^6 \text{ dynes} \cdot \text{cm}^{-2} \times 1.15 \times 10^3 \text{ cm}^3$$
$$= 3.04 \times 10^9 \text{ ergs} = 304 \text{ J},$$

which agrees with the net heat input.

Quiz: Heat and the First Law of Thermodynamics

1. Why does a marble table feel cooler to the touch than a wooden table top?

    (A) Because the marble table never absorbs enough heat to room temperature.

    (B) Because the heat conductivity of marble is higher than wood, heat flows more readily from your fingers.

    (C) Because the heat conductivity of marble is higher than wood, heat flows more readily into your fingers.

    (D) Because the heat conductivity of wood is higher than marble, heat flows more readily from your fingers.

    (E) Because the heat conductivity of wood is higher than marble, heat flows more readily into your fingers.

2. The work done in a general reversible process on an ideal gas

    (A) is independent of the process path.

    (B) depends on the process path.

    (C) depends only on its initial and final pressures.

    (D) is equal to the change in volume.

    (E) Both (B) and (C).

QUESTION 3 refers to the following specific heats of some common substances.

     I.   Lead — 0.03 calories/gram C

     II. Silver — 0.06 calories/gram C

     III. Iron — 0.11 calories/gram C

3.   Ten calories of heat are added to 1 gram of each of the substances. Which substance experiences the greatest temperature increase?

    (A) I   (D) I & II

    (B) II   (E) II & III

    (C) III

QUESTION 4 is based on the below heating curve for 10 grams of a substance.

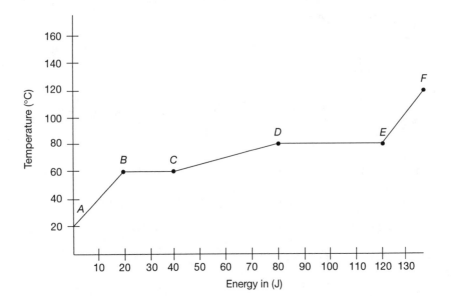

4. The heat of fusion is

   (A) 2 joules/gram.   (D) 20 joules/gram.

   (B) 4 joules/gram.   (E) 40 joules/gram.

   (C) 8 joules/gram.

QUESTION 5 is based on the heating curve for 10 grams of a substance.

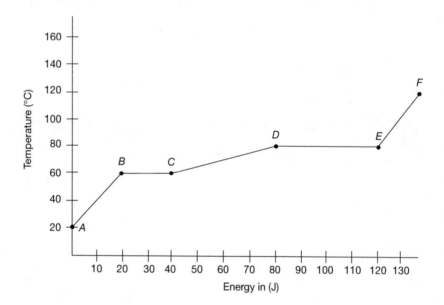

5. The heat of vaporization is

   (A) 2 joules/gram.   (D) 20 joules/gram.

   (B) 4 joules/gram.   (E) 40 joules/gram.

   (C) 8 joules/gram.

QUESTIONS 6–8 refer to the following cooling curve of a substance.

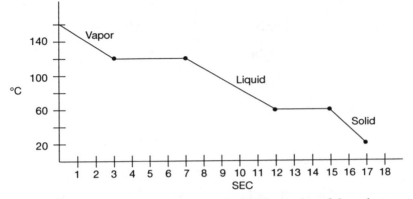

6. According to the cooling curve, the boiling point of the substance is

   (A) 0°C.   (D) 120°C.

   (B) 20°C.   (E) 160°C.

   (C) 70°C.

7. How long does it take for all of the vapor to condense to a liquid?

   (A) 3 seconds   (D) 10 seconds

   (B) 7 seconds   (E) 17 seconds

   (C) 9 seconds

8. If the specific heat of the vapor is 10 joules/gram C, how many joules of heat are given up when 2 grams of the vapor cools to the boiling point?

   (A) 20 J   (D) 800 J

   (B) 80 J   (E) 1,800 J

   (C) 400 J

QUESTIONS 9 and 10. Below is a table of specific heats for several types of matter.

Specific Heats (cal/gram C)

Aluminum	0.22
Copper	0.09
Silver	0.06
Lead	0.03
Iron	0.1

9. One hundred cals of heat is added to 10 grams of each of the metals. Which metal will have the greatest temperature increase?

   (A) Aluminum   (D) Iron

   (B) Copper   (E) Silver

   (C) Lead

10. Which metal requires the greatest number of calories to change its temperature one degree Celsius?

   (A) Aluminum   (D) Iron

   (B) Copper   (E) Silver

   (C) Lead

ANSWER KEY

1.	(B)	6.	(D)
2.	(B)	7.	(B)
3.	(A)	8.	(D)
4.	(A)	9.	(C)
5.	(B)	10.	(A)

CHAPTER 18

Kinetic Theory of Gases

18.1 Ideal Gas Equation

a)

$$pV = \mu RT$$

p = pressure (Newtons/meters^2)

V = volume (meters^2)

μ = number of moles

R = universal gas constant

T = Temperature (Kelvin)

For ideal gas: $R = 8.314$ J/mol $\bullet$ K $= 1.986$ cal/mol $\bullet$ K

b)

$$\frac{p_i V_i}{T_i} = \frac{p_f V_f}{T_f} = R = \text{a constant}$$

For a fixed mass of gas: a closed system:

i = initial value

f = final value

Problem Solving Example:

 An ideal gas is contained within a volume of 0.0566 m³, when the pressure is 137 atmosphere and the temperature is 27°C. What volume would be occupied by this gas if it were allowed to expand to atmospheric pressure at a temperature of 50°C?

 We may use the ideal gas law to analyze the behavior of the ideal gas.

$$PV = nRT$$

where P is the pressure of a container (of volume V) of gas at an absolute temperature T. n is the number of moles of the gas and R is the universal gas constant. Since we are dealing with a fixed mass of gas, we may write

$$\frac{PV}{T} = nR$$

Alternatively,

$$\frac{P_1 V_1}{T_1} = \frac{P_2 V_2}{T_2}$$

where (V_1, P_1, T_1) and (V_2, P_2, T_2) are the conditions that describe the behavior of the gas before and after it expands, respectively.

Since

$$T_1 = 273 + 27 = 300K$$

and

$$T_2 = 273 + 50 = 323K$$

then

$$V_2 = V_1 \frac{P_1}{P_2} \frac{T_2}{T_1} = \left(2 \times \frac{137}{1} \times \frac{323}{300}\right) ft^3 = 295\ ft^3$$

18.2 Pressure and Molecular Speed of an Ideal Gas

a)

$$p = \frac{1}{3} \rho v^2$$

p = pressure

ρ = density

v^2 = average square speed

b)

$$V_{rms} = \sqrt{\frac{3p}{\rho}} = \sqrt{v^2}$$

V_{rms} = root-mean-squared speed

p = pressure

ρ = density

Problem Solving Example:

 Compute the rms speed of O_2 at room temperature.

 The rms speed of a gas molecule is

$$V_{rms} = \sqrt{\frac{3RT}{M}}$$

where M is the molar mass of the gas, T is its temperature in Kelvin, and R is the gas constant. Hence,

$$V_{rms} = \sqrt{\frac{(3)(8.31 \text{ joule/moleK})(273\text{K})}{\left(32 \times 10^{-3} \text{ kg/mole}\right)}}$$

Here, we have used the fact that $0°C = (0 + 273)K$.

$$V_{rms} = \sqrt{2.13 \times 10^5 \text{ m}^2/\text{s}^2}$$

$$V_{rms} = 4.61 \times 10^5 \text{ m/s}$$

18.3 Kinetic Energy of an Ideal Gas

a)

$$\frac{1}{2} M \text{ v}^2 = \frac{3}{2} RT = \text{Kinetic Energy}$$

M = molecular weight

v^2 = average square speed

R = ideal gas constant

T = temperature

b)

$$\frac{1}{2} mv^2 = \frac{3}{2} kT$$

m = molecular mass

k = Boltzmann's Constant

T = temperature

$k = 1.381 \times 10^{-23}$ J/Molecule • K

c)

$$\frac{V_{r_1}}{V_{r_2}} = \sqrt{\frac{m_2}{m_1}}$$

V_r = root-mean-square speed

m = mass

Problem Solving Example:

 What is the average velocity of the molecules of the air at 27°C?

 From a simple atomic model in which we consider the atom as a hard spherical body subject to completely elastic collisions we can develop an equation for the average kinetic energy of a molecule which is

$$\frac{1}{2}mv_a^2 = \frac{3kT}{2}$$

where m is the mass of one molecule, v_a is its average velocity, k is Boltzmann's constant, and T is the absolute temperature of the environment of the molecule. Multiply both sides by Avogadro's number N_A and 2

$$N_A m v_a^2 = 3N_A kT$$

But $N_A m = M_m$ and $N_A k = R$ since M_m is the mass of one mole of molecules, and R is the gas constant which is Boltzmann's constant times Avogadro's number. Substituting and rearranging, we have

$$v_a^2 = \frac{3RT}{M_m}$$

Air consists mainly of nitrogen, which is diatomic, and its effective molecular mass is approximately twice the atomic mass of nitrogen. So

$$M_m \approx 2 \times 14 = 28 \text{ gm/mole}$$

$$v_a^2 = \frac{(3)(8.32 \text{ Joule/moleK})(273+27)\text{K}}{28 \text{ gm/mole}}$$

where we have used the fact that 27°C = (273 + 27)K. Hence,

$$v_a^2 = 267.43 \text{ J/gm}$$

But   1 joule = 1 nt • m = 1 kg • m²/s² = 10⁷ gm • cm²/s²

and $v_a^2 = 267.43 \times 10^7 \dfrac{\text{gm} \bullet \text{cm}^2}{\text{gm} \bullet \text{s}^2} = 267.43 \times 10^7 \text{ cm}^2/\text{s}^2$

Therefore, $v_a = 5.17 \times 10^4$ cm/s

This is equivalent to 1,160 miles per hour!

18.4 Internal Energy of an Ideal Monatomic Gas

$$U = \frac{3}{2}\mu RT$$

U = internal energy

μ = number of moles

R = universal gas constant

T = temperature

Problem Solving Example:

Q Find the minimum radius for a planet of mean density 5,500 kg • m⁻³ and temperature 400°C which has retained oxygen in its atmosphere.

The escape velocity from a planet is given by the relation

$$V = \sqrt{\frac{2GM}{r}} = \sqrt{2}\sqrt{\frac{G \times \frac{4}{3}\pi r^3 \rho}{r}} = \sqrt{\frac{8}{3}G\pi\rho r^2} \tag{1}$$

where r is the planet radius, M is its mass, and ρ is its density.

If most oxygen molecules have velocities greater than this, then, when they are traveling upward near the top of the atmosphere, they will escape into space and never return. A slow loss of oxygen from the atmosphere will therefore take place. In this case, however, we are told that the planet has retained its oxygen and we can assume that escape velocity from the planet is greater than the rms velocity of the oxygen molecules. When the two are equated, the minimum radius for the planet results. We need the rms velocity of oxygen molecules. This speed V is so defined that the internal energy U would be the same if all the atoms have this speed. For a gas consisting of N atoms, V is defined by

$$U = N\frac{1}{2}mv^2 \tag{2}$$

where m is the mass of one atom.

If the gas is ideal and monatomic, then we also know that

$$U = N\frac{3}{2}kT \tag{3}$$

where k is Boltzmann's constant and T is the temperature of the gas in Kelvin. Oxygen is neither ideal nor monatomic but equation (3) is still a good approximation since the gas is not very dense and therefore interatomic forces can be ignored.

Equating equations (2) and (3), we get

$$mv^2 = 3kT$$

Multiplying both sides by Avogadro's number N_A (the number of molecules in one mole of the gas),

$$N_A m v^2 = 3 N_A k T \qquad (4)$$

But $N_A m$ equals the mass, M', of one mole of the gas. Also, by definition $N_A k = R$, where R is the universal gas constant. Substituting these two expressions in equation (4) yields

$$M' v^2 = 3RT$$

Solving for the velocity, we have

$$v = \sqrt{\frac{3RT}{M'}} \qquad (5)$$

Set equations (1) and (5) equal to each other so as to find the minimum radius. Then

$$\sqrt{\frac{8}{3} G \pi \rho r_{\min}^2} = \sqrt{\frac{3RT}{M'}}$$

where T is the absolute temperature in the atmosphere, and M' is the mass per mole of O_2.

The temperature of the oxygen is $400°C + 273° = 673K$. Oxygen gas is diatomic, and its effective molecular mass is therefore twice the atomic mass of monatomic oxygen.

$$M' = (2 \times 16) \text{g} \cdot \text{mole}^{-1} = 32 \times 10^{-3} \text{ kg} \cdot \text{mole}^{-1}$$

$$r_{\min} = \sqrt{\frac{9RT}{8G\pi\rho M'}}$$

$$= \sqrt{\frac{9 \times 8.315 \text{ J} \cdot \text{mole}^{-1} \cdot \text{K deg}^{-1} \times 673 \text{ K deg}}{8 \times 6.67 \times 10^{-11} N \cdot \text{m}^2 \cdot \text{kg}^{-2} \times \pi \times 5,500 \text{ kg} \cdot \text{m}^{-3}}{\times 32 \times 10^{-3} \text{ kg} \cdot \text{mole}^{-1}}}$$

$$\therefore r_{\min} = \sqrt{1.708 \times 10^{11} \text{m}^2} = 4.131 \times 10^5 \text{m} = 413.1 \text{ km}.$$

18.5 Specific Heats of an Ideal Gas

a) Ideal Monatomic Gas

    i)

$$C_v = \frac{3}{2} R \approx 3 \text{ cal./moleK}$$

    C_v = molar heat capacity at constant t value

    R = universal gas constant

    ii)

$$\gamma = \frac{C_p}{C_v} \cong \frac{5}{3} \cong 1.67$$

    γ = ratio of specific heats

    C_p = molar heat capacity at constant pressure

    C_v = molar heat capacity at constant value

b) Ideal Diatomic Gas

$$\gamma = \frac{C_p}{C_v} \cong \frac{7}{5} \cong 1.40$$

c) Ideal Polyatomic Gas

$$R = C_p - C_v$$

d) Universal Gas Constant

$$R = C_p - C_v$$

Problem Solving Example:

Considering air to be an ideal gas to a first approximation, calculate the ratio of the specific heats of air, given that at sea

level and STP the velocity of sound in air is 334 m • s^{-1}, and that the molecular weight of air is 28.8 g • mole$^{-1.}$

The speed of sound in air is

$$c = \sqrt{\frac{\beta}{\rho}} \qquad (1)$$

where ρ is the density of air, and β is its bulk modulus. The latter is given by

$$\beta = \frac{-\Delta p}{(\Delta V/V)}$$

where $\Delta V/V$ is the fractional change in volume of a volume element of air when it is exposed to a change in pressure Δp. For infinitesimal increments, we may write

$$\beta = \frac{-dp}{(dV/V)} \qquad \text{or} \qquad -\frac{Vdp}{dV} \qquad (2)$$

Now, the compressions and rarefactions of the sound waves traveling through air are adiabatic. Hence, the pressure experienced by a volume of air, V, must satisfy.

$$pV^\gamma = \text{constant} = \alpha \qquad (3)$$

where $r = Cp/Cv$, the ratio of the molar specific heat at constant pressure and the molar specific heat at constant volume. Then, using (3)

$$\frac{dp}{dV} = \frac{d}{dV}\left(\frac{\alpha}{V^\gamma}\right) = \frac{d}{dV}\left(\alpha V^{-\gamma}\right)$$

$$\frac{dp}{dV} = -\gamma \, \alpha V^{-\gamma-1}$$

Since $\alpha = pV^\gamma$, this becomes

$$\frac{dp}{dV} = -\gamma \, pV^\gamma \, V^{-\gamma-1} = -\gamma pV^{-1}$$

Using this relation in (2),

$$\beta = -V\left(-\gamma p V^{-1}\right) = \gamma p$$

Inserting this in (1),

$$c = \sqrt{\frac{\gamma p}{\rho}} \tag{4}$$

But, if air is assumed to be an ideal gas, it must follow the ideal gas law, or

$$pV = \mu RT$$

where T is the temperature of the air in Kelvin, and μ is the number of moles of air in a given volume of the gas, V. Then

$$p = \frac{\mu RT}{V}$$

Now,

$$\mu = \frac{M}{M_0}$$

where M is the mass of air in a volume V, and M_0 is the mass of one mole of air. Hence,

$$p = \frac{MRT}{M_0 V} = \frac{\rho RT}{M_0}$$

by definition of ρ. Using this in (4),

$$c = \sqrt{\frac{\gamma RT}{M_0}}$$

whence

$$\gamma = \frac{M_0 c^2}{RT}$$

$$= \frac{28.8 \,\text{g} \cdot \text{mole}^{-1} \times 33,400^2 \,\text{cm}^2 \cdot \text{s}^{-2}}{8.31 \times 10^7 \,\text{ergs} \cdot \text{mole}^{-1} \cdot \text{K deg}^{-1} \times 273 \text{K deg}}$$

$$= 1.415$$

18.6 Mean Free Path

$$\ell = \frac{1}{\sqrt{2}\pi n_v d^2}$$

ℓ = mean free path

n_v = number of molecules per unit volume

d = molecular diameter

Problem Solving Example:

Show that the average distance, l, a molecule travels between collisions in a gas, is related to the number density n and the molecular diameter d by

$$\ell = \frac{1}{n\pi d^2}$$

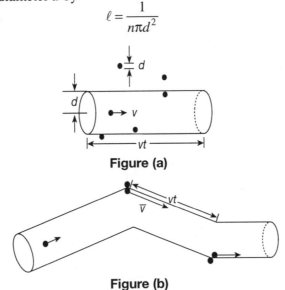

Figure (a)

Figure (b)

A   Consider a molecule moving through a region of stationary molecules. It can collide with those molecules whose centers are at a distance less than or equal to d from its center. If the speed of the molecule is v, it moves a distance vt in time t and collides with every molecule in the cylindrical volume $vt\,\pi d^2$ (Figure (a) on previous page).

The molecules in this cylinder cannot avoid the incident molecule because the diameter of the cylinder is $2d$, twice that of a molecule. If there are n molecules per unit volume, the number of molecules in the cylinder is

$$N = nvt\,\pi d^2$$

Actually, after each collision, the molecule changes its direction and the cylinder mentioned before makes zigzags as shown in Figure (b) on the previous page. The number of collisions is just the number of molecules in the cylinder; hence, the average distance per collision is

$$\ell = \frac{vt}{nvt\pi d^2} = \frac{1}{n\pi d^2}$$

e is called the mean free path.

Quiz: Kinetic Theory of Gases

1. The relationship between the pressure and volume of an ideal gas when temperature is held constant.

    (A) Logarithmic   (D) Direct

    (B) Geometric   (E) Ratio

    (C) Inverse

2. A gas at atmospheric pressure and room temperature is contained in a cylinder that is sealed on one end, and contains a frictionless movable cylinder at the other end. If the gas expands adiabatically, the piston is pushed out. If you increase the volume,

    (A) the pressure remains the same.

    (B) the temperature increases.

    (C) the temperature may increase or decrease, depending on the gas.

    (D) the temperature decreases.

    (E) there is no change in temperature.

3. According to the ideal gas law, the behavior of gases is related to the ideal gas constant, R. Thus, the dimensions of a possible value of R can be

    (A) (liter)(atm)/(mol)(°C).

    (B) (liter)(atm)/(mol)(K).

    (C) (mol)(atm)/(liter)(K).

    (D) (mol)(atm)/(liter)(°C).

    (E) (liter)(atm)/(grams)(K).

4. If you double the temperature of a mixture of H_2 and N_2 gas, which of the following will be true concerning the average kinetic energy and average velocity of each of the two gases?

   (A) Molecules of N_2 and H_2 will have the same average velocity and average kinetic energy.

   (B) Both will have the same average velocity but molecules of N_2 will have a higher average kinetic energy than molecules of H_2.

   (C) Both will have the same average velocity but molecules of H_2 will have a higher average kinetic energy than molecules of N_2.

   (D) Both will have the same average kinetic energy but molecules of N_2 will have a higher average velocity than molecules of H_2.

   (E) Both will have the same average kinetic energy but molecules of H_2 will have a higher average velocity than molecules of N_2.

5. Which of the following phenomena are impossible according to the 2nd Law of Thermodynamics?

   I. Complete conversion of heat from a single thermal reservoir into mechanical work,

   II. Reduction of temperature to absolute zero, 0K.

   III. The total entropy of an isolated system can decrease.

   IV. The total entropy of an isolated system can increase.

   (A) I and III   (D) I, II, and III

   (B) I and IV   (E) I, II and IV

   (C) II and III

6. The temperature scale used in the ideal gas law is

    (A) Fahrenheit.   (D) Celsius.

    (B) Kelvin.   (E) Metric.

    (C) English.

7. As gas is released into the atmosphere from the high pressure inside an aerosol can, it cools. This is due mainly to

    (A) the first law of thermodynamics.

    (B) the second law of thermodynamics.

    (C) the law of entropy.

    (D) Boyle's law.

    (E) the ideal gas law.

FOR QUESTIONS 8 and 9 choose the best answer from the following choices.

    I.   Isobaric

    II.   Isometric

    III. Isothermal

    IV.   Adiabatic

    V.   Carnot cycle

8. A process through which pressure is inversely proportional to volume.

    (A) I   (D) IV

    (B) II   (E) V

    (C) III

9. A process in which *P*, *V*, and *T* all change but no heat flows into or out of the system.

(A) I

(D) IV

(B) II

(E) V

(C) III

10. The tire of an automobile is filled with air to a gauge pressure of 35 psi at 20°C in the summertime. What is the gauge pressure in the tire when the temperature falls to 0°C in the wintertime? Assume that the volume does not change and that the atmospheric pressure is a constant 14.7 psi.

(A) 49.7 psi

(D) 31.6 psi

(B) 35 psi

(E) 46.3 psi

(C) 14.7 psi

ANSWER KEY

1.	(C)		6.	(B)
2.	(D)		7.	(A)
3.	(B)		8.	(C)
4.	(E)		9.	(D)
5.	(D)		10.	(D)

CHAPTER 19

Entropy and the Second Law

of Thermodynamics

19.1 Efficiency of a Heat Engine

a)

$$e = \frac{W}{Q_h} = 1 - \frac{Q_c}{Q_h}$$

e = efficiency

W = work done by engine

Q_c = heat (lower)

Q_h = heat (higher)

b)

$$e = 1 - \frac{T_c}{T_h}$$

e = efficiency

T_c = temperature (lower)

T_h = temperature (higher)

Problem Solving Example:

 Q In a certain engine, fuel is burned and the resulting heat is used to produce steam, which is then directed against the vanes of a turbine, causing it to rotate. What is the efficiency of the heat engine if the temperature of the steam striking the vanes is 400 K and the temperature of the steam as it leaves the engine is 373 K?

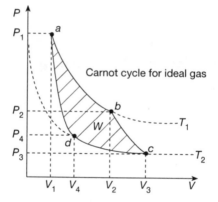

A The efficiency E of a heat engine is the ratio of the net work W done by the engine in one cycle to the heat Q, absorbed from the high temperature source in one cycle.

$$E = \frac{W}{Q_1}$$

For the Carnot cycle, which describes the operation of a reversible heat engine, we know the efficiency to be

$$E = \frac{W}{Q_1} = \frac{Q_1 - Q_2}{Q_1} = \frac{T_1 - T_2}{T_1}$$

where T_2 is the low temperature. Carnot stated that the efficiency of all Carnot engines operating between the same two temperatures is the same and that no irreversible engine working between these two temperatures can have a greater efficiency. This means that the maximum efficiency of this heat engine is given by

$$\varepsilon = \frac{T_1 - T_2}{T_1} = \frac{400\text{K} - 373\text{K}}{400\text{K}} = 0.068 = 6.8\%$$

19.2 Entropy, S

a) For a Closed Cycle,

$$\oint dS = 0$$

dS = rate of entropy change

b)

$$\int_a^b dS = \int_a^b \frac{dQ}{T} \qquad \rightarrow \text{units: Joules /Kelvin, i.e., J/K}$$

e = efficiency

T_c = temperature (lower)

T_h = temperature (higher)

c) For Free Expansion

$$S_f - S_i = \int_i^f \frac{dQ}{T} = \mu R \ln\left(\frac{V_f}{V_i}\right)$$

S_f = final entropy

S_i = initial entropy

dQ = rate of heat change

T = temperature

μ = number of moles

R = ideal gas constant

V_f = final volume

V_i = initial volume

Problem Solving Examples:

 What is the change in entropy of a gas if the temperature increases from 100K to 101K when heat is added and the volume is kept constant?

 Consider a system containing a large number of particles. When heat is added to this system, the average kinetic energy of the particles will increase. This is reflected as an increase in the temperature of the system. The system will have a higher internal disorder as a result of increased thermal motion of the constituents.

The entropy of a system is a measure of the tendency of a system to increase its internal disorder. Therefore, as heat is added, entropy increases. In our problem, let the increase in entropy be Δs when the system reaches a new equilibrium after its temperature increases by ΔT = 1 K. Since $\Delta T << T$ = 100 K, the amount of the heat added must be very small, and the entropy change is

$$\Delta s = \frac{Q}{T}$$

where Q is the quantity of heat added.

The heat added to a gas is equal to the gas's increase in internal energy plus the work done on the gas while expanding. The volume is kept constant; therefore, the mechanical work done is zero. Using N

for Avogadro's number and k for Boltzmann's constant, we have (for an ideal monatomic gas)

$$Q = \Delta E = \frac{3}{2} Nk\,\Delta T$$

where ΔE is the increase in the internal energy of the gas. Hence,

$$\Delta s = \frac{3}{2} Nk \frac{\Delta T}{T}$$
$$= \frac{3}{2} \frac{(6.02 \times 10^{23}\ \text{mole}^{-1}) \times (1.38 \times 10^{-23}\ \text{J/K})\,1 \bullet k}{100K}$$
$$= 0.125\ \text{joule/moleK}$$

 When 100 g water at 0°C are mixed with 50 g of water at 50°C, what is the change of entropy on mixing?

The 100 g of water at 0°C are arbitrarily said to have zero entropy. The 50 g of water at 50°C have a greater entropy than the same quantity of water at 0°C, since it contains more heat energy. Entropy S is defined as

$$ds = \frac{dQ}{T} \tag{1}$$

where ds is the infinitesimal change in the entropy due to an infinitesimal quantity of heat dQ in the system. T is the instantaneous temperature in Kelvin. Integrating both sides of equation (1) gives

$$S_2 - S_1 = \int_{Q_1}^{Q_2} \frac{dQ}{T} \tag{2}$$

In raising its temperature from a temperature t_1 to t_2, a substance absorbs heat dQ given by

$$dQ = mcdT \tag{3}$$

where m is the mass of the substance, c is its specific heat, and dT is its change in temperature in Kelvin (same as change in Celsius degrees). Substitution of equation (3) into equation (2) yields

$$S_2 - S_1 = \int_{t_1}^{t_2} mc \frac{dT}{T} \qquad (4)$$

In Kelvin, $0°C = (0 + 273)\ K = 273\ K$ and $50°C = (50 + 273)\ K = 323$ K. Let $m_2 = 50$ gm and $m_1 = 100$ g. The specific heat of water is given by $c = 1$ cal $\bullet$ g^{-1} $\bullet$ K deg^{-1}. Then

$$S_2 - S_1 = \int_{273K}^{323K} m_2 c \frac{dT}{T} = m_2 c \ln\left(\frac{323}{273}\right)$$

$$= 50\ g \times 1 cal \bullet g^{-1} \bullet K\, deg^{-1} \times 2.303 \times 0.0730$$

$$= 8.4\ cal \bullet K\, deg^{-1}$$

Since $S_1 = 0$, it follows that $S_2 = 8.4$ cal $\bullet$ K deg^{-1}.

When the water is mixed, the heat gained by the cold water is equal to the heat lost by the hot water. Therefore, $m_1 c\ (t_3 - t_1) = m_2 c\ (t_2 - t_3)$, where t_1 is the original temperature ($0°C$) of the 100 g of water, t_2 is the original temperature ($50°C$) of the 50 g of water, and t_3 is the final intermediate temperature of the system.

$$100\,g \times (t_3 - 0°C) = 50\,g \times (50°C - t_3)$$

$$\therefore t_3 = \frac{2,500K}{150} = 16.67°C$$

Converting to Kelvin in order to be able to use equation (4), we have $t_3 = 16.67°C = (16.67 + 273)\ K = 289.67K$.

The entropy of the final mixture is

$$S_3 = \int_{t_1}^{t_3} \frac{dQ}{T} = \int_{273\mathrm{K}}^{289.67\mathrm{K}} (m_1 + m_2) c \frac{dT}{T} = (m_1 + m_2)c \ln\left(\frac{289.67}{273}\right)$$

$$= 150 \text{ g} \times 1 \text{ cal} \bullet \text{g}^{-1} \bullet \text{K deg}^{-1} \times 2.303 \times 0.0257$$

$$= 8.9 \text{ cal} \bullet \text{K deg}^{-1}$$

The increase in entropy is thus 0.5 cal $\bullet$ K deg^{-1}.

Quiz: Entropy and the Second Law of Thermodynamics

1. What is the ideal efficiency of a steam engine that takes steam from the boiler at 200°C and exhausts it at 100°C?

    (A) 50%   (D) 11%

    (B) 25%   (E) 2%

    (C) 21%

QUESTION 2 is based on the heating curve for 10 grams of a substance shown below.

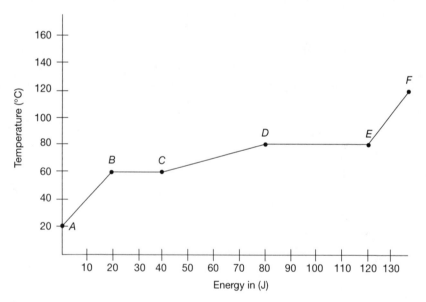

2. Entropy is increasing between

    (A) *A* and *B*.   (D) All of the above.

    (B) *B* and *C*.   (E) None of the above.

    (C) *C* and *D*.

3. Which of the following phenomena are impossible according to the second law of thermodynamics?

    I.   Complete conversion of heat from a single thermal reservoir into mechanical work.

    II. Reduction of temperature to absolute zero, OK.

    III. The total entropy of an isolated system can decrease.

    IV. The total entropy of an isolated system can increase.

    (A) I and III only.   (D) I, II, and III only.

    (B) I and IV only.   (E) I, II, and IV only.

    (C) II and III only.

4. The second law of thermodynamics, the law of entropy, would not be violated if a ball were to bounce higher and higher with each bounce and if the internal energy (temperature) dropped to compensate for its gain of kinetic energy. This unlikely event does not violate the second law of thermodynamics because

    (A) the second law is an approximate law.

    (B) the second law is not really a valid law.

    (C) the second law is a statistical law.

    (D) the second law is a misconception.

    (E) it violates the first law of thermodynamics.

5. The ratio of usable work produced by a device to the work input required by that device is called

    (A) latent heat.   (D) thermodynamics.

    (B) work output.   (E) efficiency.

    (C) capacity.

6. A process in which no heat is exchanged with the surroundings of a system is described as

    (A) isothermal.   (D) isobaric.

    (B) isometric.   (E) adiabatic.

    (C) isolated.

7. For the Carnot refrigeration cycle shown, determine the efficiency.

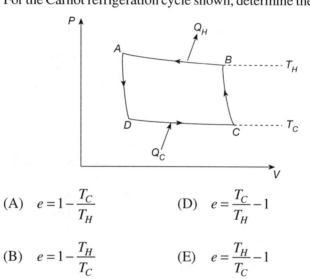

(A) $e = 1 - \dfrac{T_C}{T_H}$ (D) $e = \dfrac{T_C}{T_H} - 1$

(B) $e = 1 - \dfrac{T_H}{T_C}$ (E) $e = \dfrac{T_H}{T_C} - 1$

(C) $e = 1 + \dfrac{T_C}{T_H}$

8. The second law of thermodynamics is intimately connected with the transfer of heat and the operation of machines. Which one of the following is NOT a correct statement in light of this law?

    (A) It is not possible to *only* transfer heat into work extracted from a uniform temperature source.

    (B) It is possible to construct a perpetuum mobile of the second kind.

(C) It is impossible to *only* transfer heat from a body at high temperature to one at lower temperature.

(D) If heat flows by conduction from body A to body B, then it is impossible to *only* transfer heat from body B to body A.

(E) It is not possible to *only* transform work into heat where a body is at a uniform temperature.

9. For one mole of ideal gas and the Carnot cycle pictured below, find $Q_H - Q_C$.

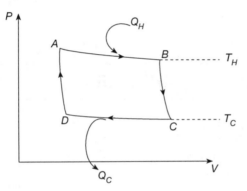

(A) $RT_H \ln \dfrac{V_A}{V_B} - RT_C \ln \dfrac{V_C}{V_D}$

(B) $RT_H \ln \dfrac{V_B}{V_A} - RT_C \ln \dfrac{V_D}{V_C}$

(C) $RT_H \ln \dfrac{V_A}{V_B} - RT_C \ln \dfrac{V_D}{V_C}$

(D) $R[T_H - T_C]$

(E) $RT_H \ln \dfrac{V_B}{V_A} - RT_C \ln \dfrac{V_C}{V_D}$

10. Consider a reversible isothermal expansion of a photon gas. Determine the entropy S for this gas at temperature T and volume V.

(A) $\sigma T^4 V$

(D) $\dfrac{2}{3}\sigma T^3 V$

(B) $\sigma T^3 V$

(E) $\dfrac{4}{3}\sigma T^3 V$

(C) $\dfrac{1}{3}\sigma T^3 V$

ANSWER KEY

1.	(C)	6.	(D)
2.	(D)	7.	(E)
3.	(D)	8.	(E)
4.	(C)	9.	(E)
5.	(E)	10.	(E)

CHAPTER 20

Electromagnetism and Electric Fields

20.1 Coulomb's Law

a) Definition of a Charge

$$q = it \quad \rightarrow \text{units: amperes} \times \text{seconds} = \text{coulombs}$$

q = electric charge

i = current

t = time

b) Coulombs Law and the Permittivity Constant, ε_0

$$F = \frac{1}{4\pi\varepsilon_0} \frac{q_1 q_2}{r^2}$$

F = force between charged particles

q_1 = electric charge of particle 1

q_2 = electric charge of particle 2

r = distance between the two centers of the particles

ε_0 = *permittivity constant

*$\varepsilon_0 = 8.854 \times 10^{-12}$ C ^{2}/N $\times m^2$

Problem Solving Example:

Two equal conducting spheres of negligible size are charged with 16.0×10^{-14} C and -6.4×10^{-14} C, respectively, and are placed 20 cm apart. They are then moved to a distance of 50 cm apart. Compare the forces between them in the two positions. The spheres are connected by a thin wire. What force does each now exert on the other?

The equation giving the force between the spheres, which may be considered as point charges, is by Coulomb's law,

$$F = \frac{1}{4\pi\varepsilon_0} \frac{q_1 q_2}{r^2}$$

where q_1 and q_2 are the charges on the spheres, and r is their separation. Thus,

$$F_1 = \frac{1}{4\pi\varepsilon_0} \frac{q_1 q_2}{(0.2)^2 \, m^2}$$

and

$$F_2 = \frac{1}{4\pi\varepsilon_0} \frac{q_1 q_2}{(0.5)^2 \, m^2}$$

Therefore,

$$\frac{F_1}{F_2} = \frac{(0.5)^2}{(0.2)^2} = 6\frac{1}{4}$$

If the spheres are joined by a wire, the charges, which are attracted to one another, can flow in the wire under the influence of the forces acting on them. The charges will neutralize as far as possible and

$$[16.0 \times 10^{-14} + (-6.4 \times 10^{-14})] = 9.6 \times 10^{-14}\,C$$

will be left distributed over the system. Neglecting the effect of the wire, by symmetry, $4.8 \times 10^{-14}\,C$ will reside on each sphere. The force between the two spheres is now

$$F = \frac{1}{4\pi\varepsilon_0}\frac{q^2}{r^2}$$

$$= 9 \times 10^9\ N \times m^2 \times C^{-2} \times \frac{(4.8 \times 10^{-14})^2 C^2}{(0.5)^2\ m^2}$$

$$= 8.29 \times 10^{-17}\ N$$

20.2 Definition of an Electric Field

$$\overline{E} = \frac{\overline{F}}{q_0}$$

$\overline{E}$ = electric field

$\overline{F}$ = electric force

q_0 = positive test charge

Problem Solving Example:

What is the electric field intensity at a point 30 centimeters from a charge of 0.10 coulombs?

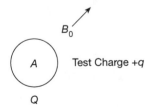

The electrostatic force on the test charge q, due to the charge Q, is, by Coulomb's law (see figure on previous page),

$$F = \frac{kQq}{r^2}$$

The electric field intensity at point B is defined as:

$$\overline{E} = \frac{\overline{F}}{q} = \left(\frac{kQq / r^2}{q} \right)$$

$$= \frac{kQ}{r^2}$$

In the problem we are presented with, we have

$Q = 0.10$ coulomb,

$r = 30$ cm $= 3.0$ m $\times 10^{-1}$ m

Then

$$\overline{E} = 9 \times 10^9 \, \frac{N - m^2}{\text{coulombs}^2} \times \frac{0.10 \ \text{coulombs}}{\left(3.0 \times 10^{-1} \right)^2 m^2}$$

$$= 10^{10} \ \text{Newton} / \text{Coulombs}$$

20.3 Lines of Force

a) For a Negatively Charged Particle

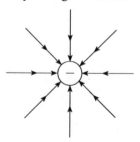

b) For a Positively Charged Particle

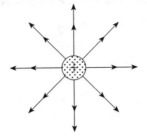

c) Like Charges Repel

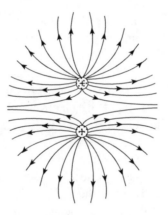

d) Opposite Charges Attract

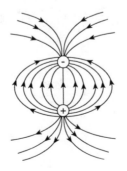

Problem Solving Example:

 Q Show how two metallic balls that are mounted on insulating glass stands may be electrostatically charged with equal amounts but opposite sign charges.

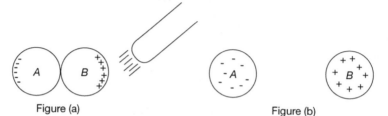

Figure (a)   Figure (b)

A The two metal balls are assumed to be initially uncharged and touching each other. (Any charge on them may first be removed by touching them to the earth. This will provide a path for the charge on the spheres to move to the ground.) A charged piece of amber is brought near one of the balls (b), as shown in the figure. The negative charge of the amber will repel the electrons in the metal and cause them to move to the far side of A, leaving B charged positively. If the balls are now separated, A retains a negative charge and B has an equal amount of positive charge. This method of charging is called charging by induction, because it was not necessary to touch the objects being electrified with a charged object (the amber). The charge distribution is induced by the electrical forces associated with the excess electrons present on the surface of the amber.

20.4 Electric Field for a Point Charge

$$E = \frac{1}{4\pi\varepsilon_0} \frac{q}{r^2}$$

E = electric field

q = point charge

r = distance from a test charge, q_0, to the point charge, q

ε_0 = permittivity constant

Problem Solving Example:

What is the intensity of the electric field 10 cm from a negative point charge of 500 stat-coulombs in air?

The electrostatic force on a positive test charge q' at a distance r from a charge Q is, by Coulomb's law (in the CGS system of units),

$$\overline{F} = k\frac{Qq'}{r^2}$$

The electric field intensity E is defined as the force per unit charge, or

$$\overline{E} = \frac{\overline{F}}{q'} = \frac{kQ}{r^2}$$

$\overline{E}$ points in the direction the force on the test charge acts. In a vacuum in cgs, $k = 1$. Therefore, the electric field 10 cm from a point charge of 500 stat-coulomb is

$$\overline{E} = \frac{500 \text{ stat-coul}}{\left(10 \text{ cm}\right)^2} = 5 \text{ dyne/stat-coul}$$

pointing directly toward the negative charge.

20.5 Torque and Potential Energy of a Dipole

a) Electric Dipole in a Uniform External Electric Field

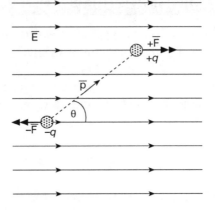

b) Torque on a Dipole

$$\overline{\tau} = \overline{p} \bullet \overline{E}$$

$\overline{\tau}$ = torque

$\overline{p}$ = dipole moment of point charges

$\overline{E}$ = external electric field

c) Potential energy of a dipole

$$U = -\overline{p} \bullet \overline{E}$$

U = potential energy

$\overline{p}$ = dipole moment of point charges

$\overline{E}$ = external electric field

CHAPTER **21**

Gauss' Law

21.1 Flux of the Electric Field

$$\phi_E = \int \overline{E} \cdot d\overline{S}$$

ϕ_E = flux of an electric field

$\overline{E}$ = electric field

S = surface area

Problem Solving Example:

The nucleus of an atom has a charge $+2e$, where e is the electronic charge. Find the electric flux through a sphere of radius 1 Å (10^{-10} m). (See figure on the following page.)

This problem is solved most directly by using Gauss' Law:

$$\Phi_E = 4\pi k_E q$$

where q is the total charge enclosed by the sphere. The orientation of the charge within the sphere does not matter. Gauss' Law yields:

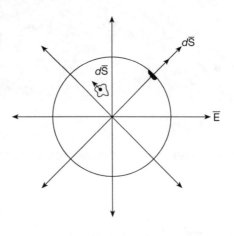

$$\Phi_E = 4\pi k_E (2e) = 8\pi k_E\, e$$

We can also find the flux from its definition:

$$\Phi_E = \int \overline{E} \cdot d\overline{S}$$

where $d\,\overline{S}$ is the differential surface area vector which is always perpendicular to the surface. Taking the nucleus to be at the center of the sphere, we see that the field, being radial, is perpendicular to the sphere surface. Since the nucleus is equidistant from all points on the sphere, the magnitude of the field is constant over the entire surface, and is, by definition:

$$E = k_E \frac{2e}{R^2}$$

where R is the radius of the sphere.

Thus, we may take $\overline{E}$ outside of the integral sign:

$$\Phi_E = 2k_E \frac{e}{R^2} \int \cos\ 0°\ dS = 2k_E \frac{e}{R^2} \int dS$$

$$= 2k_E \frac{e}{R^2} S = 2k_E \frac{e}{R^2} \times 4\pi R^2 = 8\pi k_E e$$

The $\overline{E}$ vector makes an angle of $0°$ with the $d\overline{S}$ vector, since both are perpendicular to the surface of the sphere (see diagram on previous page).

The cosine is introduced since the integrand is the inner product of the two vectors $\overline{E}$ and $d\overline{S}$, and by definition, if $\overline{A}$ and $\overline{B}$ are vectors then:

$$\overline{A} \cdot \overline{B} = |\overline{A}| \, |\overline{B}| \, \cos (A, B)$$

where (A, B) is the angle between $\overline{A}$ and $\overline{B}$.

$S = 4\pi R^2$ is the surface area of the sphere.

Substituting

$$k_E = 9 \times 10^9 \, \frac{N \cdot m^2}{coul^2} \quad \text{and } e = 1.6 \times 10^{-19} \text{ coulombs}$$

we obtain the magnitude of the flux.

$$\Phi_E = (8)(3.14)\left(9 \times 10^9 \, \frac{N \cdot m^2}{coul^2} \right)\left(1.6 \times 10^{-19} coul\right)$$

$$= 3.61 \times 10^{-8} \, \frac{N \cdot m^2}{coul}$$

We note that this result is independent of the radius of the sphere. We could have seen this immediately from Gauss' Law, since the flux through the sphere is entirely determined by the amount of charge enclosed within it.

21.2 Gauss' Law

$$q = \varepsilon_0 \int \overline{E} \cdot d\overline{S}$$

q = total charge within the surface

ε_0 = permittivity constant

$\overline{E}$ = electric field

$\overline{S}$ = surface area

Problem Solving Example:

 Consider a sphere of radius a which has a charge q evenly distributed on its surface. What is the electric field outside the sphere?

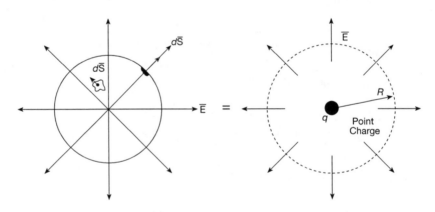

Due to the fact that the charge is uniformly distributed over the surface of the sphere (and therefore symmetric), we realize that the electric field must have the same strength at any point a distance R away from the center of the charged sphere. Furthermore, we expect the lines of $\overline{E}$ to begin on the positive surface charges and emanate radially from the surface of the sphere. Consider a spherical surface of radius R and with the same center as the charged sphere. If the field at this surface is E, then the total electric flux through the surface is

$$\Phi_E = \int \overline{E} \cdot d\overline{S} = \int E \cdot dS$$

because $\overline{E}$ and $d\overline{S}$ are parallel. Hence,

$$\Phi_E = 4\pi R^2 E$$

where $4\pi R^2$ is the surface area of the spherical surface. Since the charge enclosed by the surface is q, then by Gauss' Law,

$$\Phi_E = \frac{q}{\varepsilon_0} .$$

Therefore,

$$4\pi R^2 E = \frac{q}{\varepsilon_0}$$

from which

$$E = \frac{1}{4\pi\varepsilon_0} \frac{q}{R^2}$$

$$= k_E \frac{q}{R^2}$$

Thus, the field outside the sphere is just as if all the charge q were located at the center of the sphere.

21.3 Applications of Gauss' Law

a) Electric field inside a spherical charge

$$E = \frac{1}{4\pi\varepsilon_0} \frac{q}{r^2}$$

Here q is the portion of the total charge of the sphere that is located within radius r.

b) Electric field inside a uniform spherical charge

$$E = \frac{1}{4\pi\varepsilon_0} \frac{qr}{R^3}$$

Here r is the distance of the point charge as measured from that point to the center of the spherical charge distribution, and R is the radius of the spherical charge distribution.

c) Electric field due to an infinite line of charge

$$E = \frac{\lambda}{2\pi\varepsilon_0 r}$$

Here λ is the change per unit length (linear charge density), and r is the radius of the charge distribution.

d) Electric field due to an infinite sheet of charge

$$E = \frac{\sigma}{2\varepsilon_0}$$

Here σ is the surface charge density (change per unit area).

e) Electric field near a charged conductor

$$E = \frac{\sigma}{\varepsilon_0}$$

Problem Solving Examples:

Q By direct calculation, determine the value of the electric intensity at any distance from an infinite plane sheet of uniformly distributed charge. Show that the result follows at once from an application of Gauss' law. (See figures on the next page.)

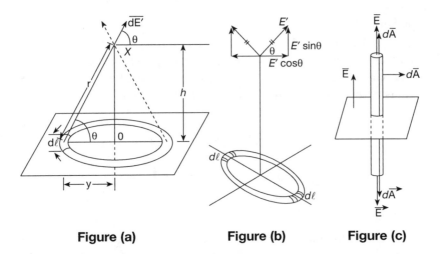

Figure (a)   **Figure (b)**   **Figure (c)**

A   Consider any point X at a perpendicular distance h from the plane sheet of charge density ρ. (See Figure (a)). Drop the perpendicular from the point to the sheet, cutting the latter at 0, and draw two circles, centered at 0, at radii of y and $y + dy$. Take a small portion of the annulus so formed, of length d ℓ, and consider the electric intensity dE' at the point X due to the small (almost point-like) element of charge. Then the electric field due to a point charge is

$$dE' = \frac{dq}{4\pi\varepsilon_0 r^2}$$

Since the charge density $\quad \rho = \frac{dq}{dA} = \frac{dq}{d\ell dy}$

then, $\qquad\qquad dE' = \frac{\rho d\ell dy}{4\pi\varepsilon_0 r^2}.$

The direction of dE' is the same as that of r, and dE' may therefore be resolved into components along OX and at right angles to it. The component along OX has the same value; no matter what position on the annulus d ℓ occupies. But the element of the annulus diametrically opposite d ℓ produces a component perpendicular to *OX* equal but opposite to that produced by d ℓ. (See Figure (b) on previous page.) These two components thus cancel out, as do all components from diametrically opposite elements. The electric intensity from the whole annulus is thus perpendicular to the sheet and has magnitude

$$dE = \oint dE' \sin\theta = \frac{\rho dy}{4\pi\varepsilon_0 r^2} \sin\theta \oint d\ell$$

$$= \frac{\rho dy}{4\pi\varepsilon_0 r^2} \times \frac{h}{r} \, 2\pi y = \frac{h\rho y \, dy}{2\varepsilon_0 \left(h^2 + y^2\right)^{3/2}}$$

We used the fact that $\sin\theta = h/r$ (geometric considerations in Figure (a)) and that the sum of all the infinitesimal elements of length d ℓ about the whole ring is equal to the circumference of the ring. Also used was the fact that $r = (h^2 + y^2)^{1/2}$.

For the whole sheet of charge, the electric intensity is the sum of contributions due to all the annuli of radius $y = 0$ to $y = \infty$ (for the infinite plane sheet). Or

$$E = \int dE = \int_0^\infty \frac{h\rho y dy}{2\varepsilon_0 \left(h^2 + y^2\right)^{3/2}} = \frac{h\rho}{2\varepsilon_0} \int_0^\infty \frac{y dy}{\left(h^2 + y^2\right)^{3/2}}$$

$$= -\frac{h\rho}{2\varepsilon_0} \left[\left(h^2 + y^2\right)^{-1/2}\right]_0^\infty$$

$$= -\frac{h\rho}{2\varepsilon_0} \left[0 - \frac{1}{h}\right] = \frac{\rho}{2\varepsilon_0}.$$

To apply Gauss' law to the same problem, construct a cylinder of small and uniform cross-sectional area *dA* at right angles to the sheet and bisected by the sheet (see Figure (c) on previous page). Since the sheet is infinite and the charge uniformly distributed, the electric intensity must be the same at all points equidistant from the sheet, and thus by symmetry must be everywhere perpendicular to the sheet. Hence, E is everywhere parallel to the sides of the cylinder and thus the flux of E from the cylinder through its sides is zero. The magnitude of E at each end of the cylinder will be the same if the cylinder is bisected by the sheet, and E will be perpendicular to each end. Hence, applying Gauss' law, we obtain

$$\int \overline{E} \cdot d\overline{A} = \int \left((\overline{E} \cdot d\overline{A})_{top} + (\overline{E} \cdot d\overline{A})_{side} + (\overline{E} \cdot d\overline{A})_{bottom} \right)$$

Since $d\overline{A}$ is small,

$$\int \overline{E} \cdot d\overline{A} \approx \overline{E} \cdot d\overline{A}$$

and

$$\int \overline{E} \cdot d\overline{A} = (EdA + 0 + EdA = 2EdA) = \frac{dq}{\varepsilon_0}$$

but

$$\rho = \frac{dq}{dA}$$

and

$$2EdA = \rho dA / \varepsilon_0.$$

Therefore, $E = \rho/2\varepsilon_0$. Thus, E is everywhere perpendicular to the sheet and has the same value $\rho/2\varepsilon_0$ at all points.

Q A ring of charge with radius 0.5 m has a 0.02 m gap (see Figure (a)). Compute the field at the center if the ring carries a charge of +1 coulomb.

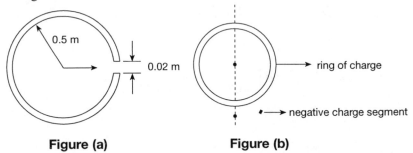

Figure (a)   **Figure (b)**

0.5 m

0.02 m

ring of charge

negative charge segment

A The field can be found by superposition of the fields of an imaginary ring of charge and a negative charge segment located where the gap would be (see Figure (b) on previous page).

We must first calculate the linear charge density of the incomplete ring.

$$\sigma = \frac{q}{\text{length of ring}} = \frac{q}{2\pi R - 0.02}$$

$$= \frac{1 \text{ coulomb}}{[2\pi(0.5) - 0.02] \text{ m}}$$

$$= 0.320 \text{ coulombs/m}$$

where σ is the linear charge density and $R = 0.5$ is the radius. We can approximate the arc length of the gap with the linear distance between the ends since this chordal length is small compared with the radius.

The field at the center due to the complete ring is zero. Since all charge elements are diametrically opposite to the center, all field elements must cancel. Thus, the field is entirely due to the negative charge segment. Both the ring and the segment must have charge densities whose absolute values equal the charge density of the incomplete ring.

$$|\lambda'| = |\delta| = -0.320 \text{ c/m}$$
$$\lambda' = -0.320 \text{ c/m} = q'/d$$

where $d = 0.02$ m

Thus, the charge on the segment is:

$$q' = d\lambda' = (0.02 \text{ m})(-0.320 \text{ c/m}) = -6.41 \times 10^{-3} \text{C}$$

The field at the center is:

$$E = k\frac{q'}{r^2} = 9 \times 10^9 \frac{\text{N m}^2}{\text{C}^2}\left(\frac{(-6.41 \times 10^{-3}\text{C})}{(0.5 \text{ m})^2}\right)$$

$$= 2.31 \times 10^8 \text{ N/C}$$

where we have treated the segment as if it were a point source.

![Q] Two hollow spherical shells are mounted concentrically, but are insulated from one another. The inner shell has a charge Q and the outer shell is grounded. What is the electric intensity and potential in the space between them?

When the outer shell is not grounded, why does a charge outside the system experience a force when the inner shell is charged?

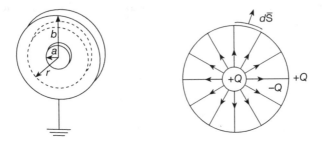

![A] In the region between the shells, because of the symmetry of the arrangement, the electric intensity $\overline{E}$ must have the same magnitude at all points a distance r from the common center, and must thus be everywhere radial in direction. Hence, applying Gauss' law to a region bounded by a spherical surface of radius r (see the figure), we have

$$\int \overline{E} \cdot d\overline{S} = \frac{Q}{\varepsilon_0} \qquad (1)$$

where Q is the total charge enclosed by the Gaussian surface, and $d\overline{S}$ is an element of surface area of the Gaussian surface. $\overline{E}$ is parallel to $d\overline{S}$ thus, (1) becomes

$$E \int dS = 4\pi r^2 E = \frac{Q}{\varepsilon_0}$$

or
$$E = \frac{Q}{4\pi\varepsilon_0 r^2} \ .$$

But $E = -dV/dr$ by definition of E as a potential gradient.

$$\therefore \quad \int dV = -\frac{Q}{4\pi\varepsilon_0 r^2} \, dr$$

or

$$V = \frac{Q}{4\pi\varepsilon_0 r} + C,$$

where C is a constant of integration. But at the outer spherical shell of radius $r = b$, $V = 0$.

$$\therefore \quad 0 = \frac{Q}{4\pi\varepsilon_0 b} + C$$

or

$$C = -\frac{Q}{4\pi\varepsilon_0 b}$$

$$\therefore \quad V = \frac{Q}{4\pi\varepsilon_0} \left(\frac{1}{r} - \frac{1}{b} \right)$$

Lines of force come from the inner shell and all end on the outer shell. The same number of lines of force end on the outer shell as start on the inner shell; hence, the charge induced on the inside of the outer shell is equal and opposite to that on the inner shell. Thus, a charge $-Q$ is induced on the inside of the outer shell. But this shell was initially uncharged. Hence, a charge $+Q$ must be left on the outside of the outer shell.

If the outer shell is grounded, electrons flow from the earth to neutralize this positive charge. In the absence of grounding, the positive charge remains on the outside of the shell and produces a field of force around the system that affects any other charge in the vicinity.

21.4 Various Gaussian Surfaces

a) Spherically Symmetric Charge Distribution

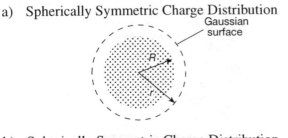

b) Spherically Symmetric Charge Distribution

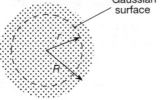

c) Infinite Line of Charge Inside a Closed Coaxial Cylindrical Gaussian Surface

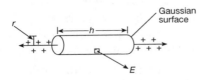

d) An Infinite Sheet of Charge Pierced by a Cylindrical Gaussian Surface

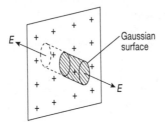

e) A Charged Insulated Conductor with a Gaussian Surface

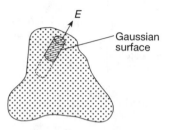

Problem Solving Example:

A charge is placed on a string of infinite length so that the linear charge density on the string is n coulombs/m. Find the electric field due to this charge distribution.

This problem can be solved using Gauss' law.

We construct as our Gaussian surface a cylinder, whose axis of symmetry coincides with the string, of length L and of radius R. It can be seen from symmetry that there can be no component of the electric field parallel to the string, since all contributions to the field in that direction will cancel (see Figure (b) on the next page). Since the field vector can be expressed in terms of components that are either perpendicular or parallel to the string, the resulting field will be radial. It can also be seen from symmetry that the magnitude of the field will be uniform over the surface of the cylinder, excluding the circular top and bottom. The flux through these portions is zero, however, since the field lines do not pass through their area.

The flux through the cylinder is therefore:

$$\phi_E = \int \overline{E} \cdot d\overline{S} = \overline{E} \int \cos\ 0°\, dS = \overline{E} \int dS$$
$$= E(2\pi RL)$$

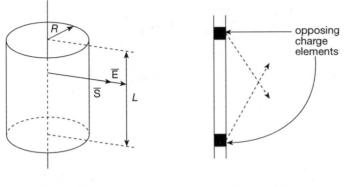

Figure (a)   **Figure (b)**

Since the field is constant in magnitude, it can be factored outside of the integral sign. The field lines coincide with the surface vector elements of the cylinder, which is the same as saying that they make an angle of 0° with each other. The corresponding term cos 0° in the integral reduces to 1, leaving only dS in the integrand which reduces to S, the total surface area of the cylinder (excluding the top and bottom). We were given:

$$\lambda = n = q/L$$

Therefore the charge on length L of string is nL, thus by Gauss' law:

$$\phi_E = 4\pi k_E q = 4\pi \frac{1}{4\pi\varepsilon_0} q = \frac{q}{\varepsilon_0}$$

$$E \cdot 2\pi R L = \frac{nL}{\varepsilon_0}$$

$$E = \frac{n}{2\pi\varepsilon_0 R} = 2k_E \frac{n}{R}$$

where

$$k_E = \frac{1}{4\pi\varepsilon_0}.$$

Quiz: Electromagnetism— Gauss' Law

QUESTIONS 1–3 refer to the following diagram.

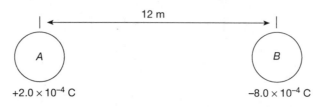

1. According to the diagram, what is the electrostatic force exerted on sphere *A*?

$$k = 9 \times 10^{9} \, \text{N m}^2/\text{C}^2$$

   (A) 1.1×10^{-9} N   (D) 10 N

   (B) 1.3×10^{-8} N   (E) 100 N

   (C) 120 N

2. Compared to the force exerted on *B* at a separation of 12 meters, the force exerted on sphere *B* at a separation of 6 meters would be

   (A) $\dfrac{1}{2}$ as great.   (D) 4 times as great.

   (B) 2 times as great.   (E) 9 times as great.

   (C) $\dfrac{1}{4}$ as great.

3. If the two spheres are touched and then separated, the charge on sphere *A* would be

   (A) -6.0×10^{-4} C.   (B) 2.0×10^{-4} C.

(C) -3.0×10^{-4} C. (D) -8.0×10^{-4} C.

(E) 8.0×10^{-4} C.

4. Two small charged pith balls separated by a distance D repel each other with a force $\bar{\bar{F}}$. If the distance between them is doubled, the force of repulsion will be

(A) $\dfrac{1}{4}$ $\bar{\bar{F}}$. (D) 1.4 $\bar{\bar{F}}$.

(B) $\dfrac{1}{2}$ $\bar{\bar{F}}$. (E) 4 $\bar{\bar{F}}$.

(C) $.707$ $\bar{\bar{F}}$.

5. Which one of the following diagrams best represents the most intense electric field created by two opposite charges?

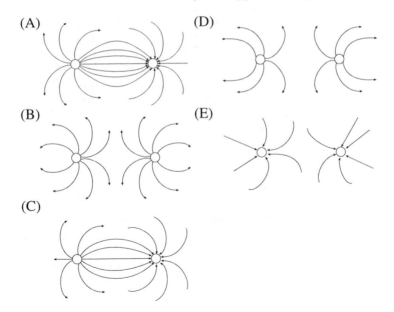

6. If a negatively charged rod is held near an uncharged metal ball, the metal ball

    (A) becomes negatively charged.

    (B) becomes positively charged.

    (C) becomes polar.

    (D) is unaffected.

    (E) Effect cannot be determined.

7. The electric field at a point near a charged object, P, is

    (A) the force on a unit negative charge at that point.

    (B) the force on a unit positive charge at that point.

    (C) the direction of the electric potential at that point.

    (D) the acceleration of an electron at that point.

    (E) the energy contained in the field at that point.

8. An electroscope is positively charged. If a glass rod that has been rubbed with silk is brought near the electroscope bulb,

    (A) the leaves will diverge.

    (B) the leaves will collapse.

    (C) nothing will happen.

    (D) the leaves will collapse then diverge.

    (E) None of the above.

9. A parallel plate capacitor is connected to a fixed voltage. If a dielectric material is introduced between the plates, which of the following will result?

    (A) Charge decreases and capacitance increases

    (B) Charge and capacitance decrease

    (C) Charge increases and capacitance decreases

    (D) Charge and capacitance increase

    (E) Charge and capacitance are not changed

10. The conventional representation of electric field lines surrounding a negatively charged object shows arrows

    (A) surrounding a charged object.

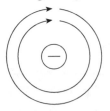

    (B) emanating from the poles of the charged object.

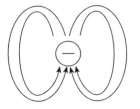

    (C) in the direction to the nearest positive charge.

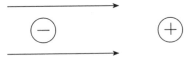

(D)   emanating radially outward from the charge.

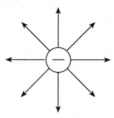

(E)   emanating radially inward toward the charge.

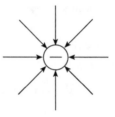

ANSWER KEY

1.	(D)	6.	(C)
2.	(D)	7.	(B)
3.	(C)	8.	(A)
4.	(A)	9.	(D)
5.	(A)	10.	(E)

CHAPTER 22

Electric Potential (Voltage)

22.1 Electric Potential

a) Electric Potential Difference

$$\Delta V = V_B - V_A = \frac{-W_{AB}}{q_0} \rightarrow \text{units: volts}$$

ΔV = potential difference

$\phi(B) = V_B$ = electric potential at point B

$\phi(A) = V_A$ = electric potential at point A

W_{AB} = work done by external force

q_0 = electrical test charge

b) Electrical Potential at a Point

$$V_B = \frac{-W}{q_0} \rightarrow \text{units: Volts}$$

This occurs when the potential drop is over an infinitesimal distance: $V_A = 0$.

Problem Solving Examples:

 a) What is the magnitude of the electric field at a distance of 1 Å (= 10^{-8} cm) from a proton? b) What is the potential at this point? c) What is the potential difference, in volts, between positions 1 and 0.2 Å from a proton?

a) From Coulomb's law

$$E = \frac{e}{r^2} \approx \frac{5 \times 10^{-10} \, \text{statcoulomb}}{\left(1 \times 10^{-8} \text{cm}\right)^2} \approx 5 \times 10^6 \, \text{statvolts/cm}$$

$$\approx (300)\left(5 \times 10^6\right) \text{V} / \text{cm} \approx 1.5 \times 10^9 \, \text{V/cm}$$

Here, e is the unit of electronic charge, and r is the distance between the proton and the point at which we calculate the field. We have also used the fact that

$$\frac{1 \, \text{statvolt}}{\text{cm}} = 300 \frac{\text{V}}{\text{cm}}$$

The field is directed radially outward from the proton.

b) The electrostatic potential at a distance r from the proton is

$$\Phi(r) = \frac{e}{r} \approx \frac{5 \times 10^{-10} \, \text{statcoulomb}}{1 \times 10^{-8} \text{cm}}$$

$$\approx 5 \times 10^{-2} \, \text{statvolts} \approx 15 \text{ V}$$

from the conversion factor given above.

c) The potential at 1×10^{-8} cm is 15 V; at 0.2×10^{-8} cm it is 75 V. The difference $75 - 15 = 60$ V.

Q In the simple case of the field due to a single point charge q, check the method for obtaining E from Φ. See the figure.

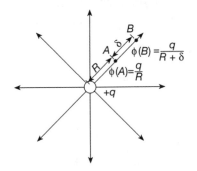

A Choose the points A and B to be on the same radius, in which case the component of E in the direction $\overline{AB}$ is the total field. Let A be at a distance R from q and B at a distance $(R + \delta)$, where δ is very small. In general,

$$\text{work} = \int \overline{F} \cdot d\overline{r}$$

where the integral is evaluated over the path of travel of the object which we are working on. For this problem, $\overline{F}$ and $d\overline{r}$ are parallel because we are moving q along a radius. Hence,

$$\text{work} = \int_{R}^{S+R} F \, dr$$

Now, because δ is very small, F will be essentially constant along the path of travel, and we may replace it by its average value, $\overline{F}$. Therefore,

$$\text{work} = \overline{F} \int_{R}^{S+R} dr \tag{1}$$

$$= \overline{F}(\delta + R - R)$$

$$= \overline{F}\delta$$

(This relation becomes more exact as δ gets smaller.) Also note that, by definition of the potential difference, between R and $R + \delta$, we obtain

$$\text{work} = {}^{-}q\Delta V$$

Substituting this in (1), we find

$$^{-}q\Delta V = \overline{F}\delta$$

But $$\overline{F} = q\overline{E}$$

Hence, $$^{-}\Delta V = \overline{E}\delta$$

or $$\overline{E} = \frac{^{-}\Delta V}{\delta} \tag{2}$$

As $\delta \to 0$, we will obtain E, the exact value of the electric field intensity.

The potential at A is by definition

$$\phi(A) = \frac{q}{R}$$

The potential at B is

$$\phi(B) = \frac{q}{R+\delta}$$

Using equation (2),

$$\overline{E} = \frac{\dfrac{q}{R} - \dfrac{q}{R+\delta}}{\delta}$$

$$= \frac{1}{\delta}\left(\frac{q(R+\delta)}{R(R+\delta)} - \frac{qR}{(R+\delta)R} \right)$$

$$= \frac{1}{\delta}\frac{(qR + q\delta - qR)}{R(R+\delta)}$$

$$= \frac{q}{R(R+\delta)}$$

$$= \frac{q}{R^2\left(1+\dfrac{\delta}{R}\right)}$$

Taking the limit as $\delta \to 0$, we find

$$\overline{E} = \frac{q}{R^2}$$

for the value of E at a distance R from q.

22.2 Potential Difference in a Uniform Electric Field

$$V_B - V_A = \frac{-W_{AB}}{q_0} = -\overline{E}d \quad \to \text{ units: volts}$$

Here $\overline{E}$ is a uniform electric field, and d is the distance between points A and B along $\overline{E}$.

Integral formula for potential difference of a known uniform field, E

$$V = -\int_{\alpha}^{\beta} \overline{E} \cdot d\ell \quad \to \text{ units: volts}$$

V = electric potential difference

α to β = interval of integration

$d\ell$ = differential displacement

$\overline{E}$ = electric field

Problem Solving Examples:

If the electrode spacing is $1 \times 10^{-3}\,m$ and the voltage is $1 \times 10^{4}\,V$, how large is the Coulomb force that is responsible for the separation of dust particles from the air? Assume that the particle's charge is + 200 quanta.

The electric field E existing between two plates is

$$E = \frac{V}{d} = \frac{1 \times 10^{4}\ \text{V}}{1 \times 10^{-3}\ \text{m}} = 1.0 \times 10^{7}\ \text{V/m}$$

Quanta represent the number of unit charges present on the dust particle. A unit charge is the charge present on a proton or 1.6×10^{-19} coulombs. Therefore, the electrostatic force experienced by the particle is

$$F = qE = (2 \times 10^{2})(1.6 \times 10^{-19}\,\text{C})(1 \times 10^{7}\,\text{V/m})$$
$$= 3.2 \times 10^{-10}\,\text{N}$$

This is approximately 10 to 100 times larger than the gravitational force on the particle.

If a 1,000-V battery is connected to two parallel plates separated by 1 mm (10^{-3} m), what is the electric field?

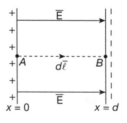

The potential of the battery, $V = 1,000$ V, and the distance between the plates, $d = 10^{-3}$ m, are given. By definition, the difference in potential experienced by moving a charge from point A to point B is

$$V_B - V_A = -\int_A^B \overline{E} \cdot \overline{d\ell} \qquad (1)$$

where E is the electric field and d is an element of the path traversed in moving the charge. Now, looking at the figure on the previous page, we see that, for the plates of a battery, E is perpendicular to the plates. If we evaluate (1) over a straight line path parallel to E, we find

$$V_B - V_A = -\int_A^B E \cdot d\ell = -\int_0^d E d\ell = -Ed$$

where d is the plate separation. Then

$$|E| = \frac{|V_b - V_a|}{d} = \frac{10^3 \, \text{V}}{10^{-3} \, \text{m}} = 10^6 \, \text{V/m}.$$

22.3 Electric Potential Due to a Point Charge

$$V = \frac{1}{4\pi\varepsilon_0} \frac{q}{r}$$

V = electric potential

q = point charge

r = distance from test charge to q

ε_0 = permittivity constant

22.4 Potential of a Group of Point Charges

a) Electric Potential Due to a Group of Point Charges

$$V = \sum_n V_n = \frac{1}{4\pi\varepsilon_0} \sum_n \frac{q_n}{r_n}$$

V = total electric potential

V_n = electric potential due to each charge

q_n = value of nth charge, and

r_n = distance of this charge from an arbitrary reference point

b) Electric Potential Due to a Continuous Charge Distribution

$$V = \int dV = \frac{1}{4\pi\varepsilon_0} \int \frac{dq}{r}$$

V = electric potential

dq = differential charge

dV = differential potential difference

r = distance of q from an arbitrary reference point

Problem Solving Example:

Compute the electric field and the electric potential at point P midway between two charges, $q_1 = q_2 = +5$ statC, separated by 1 m.

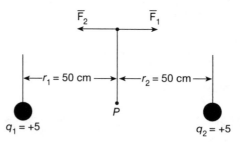

The magnitude of a test charge is +1 unit of charge. The forces on a test charge placed midway between two identical charges q_1 and q_2 are

$$\bar{F}_1 = \frac{q_1}{r_1^2} \hat{r}_{12}$$

Electric Potential Due to a Dipole

$$\overline{F}_2 = \frac{q_2}{r_2^2}\hat{r}_{21} = \frac{-q_2}{r_2^2}\hat{r}_{12} = \frac{-q_1}{r_1^2}\hat{r}_{12}$$

where the unit vector $\hat{r}_{12}$ points from q_1 to q_2, r_1, r_2 are the distances between the test charge and q_1 and q_2, and $\overline{F}_1$ and $\overline{F}_2$ are equal in magnitude but opposite in direction. The net force is therefore zero at P;

$$F = F_1 + F_2 = 0.$$

The force on a unit charge gives the electric field strength at that point; then the electric field $\overline{E}$ at P is also zero.

Although the electric field at P is zero, this does not imply that the electric potential is also zero. The total potential V_{total} is the sum (the algebraic sum since potential is a scalar) of the potentials due to q_1 and q_2:

$$\Phi_{E,1} = V_{E,1} = \frac{q_1}{r_1} = \frac{5}{50} = 0.1 \text{ statV}$$

$$\Phi_{E,2} = V_{E,2} = \frac{q_2}{r_2} = \frac{5}{50} = 0.1 \text{ statV}$$

Therefore, $\qquad V_{E,total} = \Phi_{E,1} + \Phi_{E,2} = 0.2 \text{ statV}$

Notice that if either q_1 or q_2 is changed from +5 statC to –5 statC, the electric potential will vanish but the electric field will not. Therefore, the fact that either the field or the potential is zero in any particular case does not necessarily mean that the other quantity will also be zero; each quantity must be calculated separately.

22.5 Electric Potential Due to a Dipole

$$V = \frac{1}{4\pi\varepsilon_0}\frac{p\cos\theta}{r^2}$$

where $(q)(2a) = p = $ dipole moment

$$V = \frac{q}{4\pi\varepsilon_0} \frac{2a\cos\theta}{r^2}$$

V = Electric Potential

a = Distance Between the Center of Dipole and Point Charge

q = Magnitude of Point Charge

r = Distance from the center of Dipole to Point in the field of Dipole

θ = angle between dipole axis and the line through its center as drawn from the point in the dipole field

Problem Solving Example:

 Show that, for a given dipole, V and E cannot have the same magnitude in MKS units at distances less than 2 m from the dipole. Suppose that the distance is $\sqrt{5}$ m; determine the directions along which V and E are equal in magnitude.

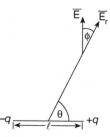

The expression for the magnitude of the potential due to a di pole is

$$V = \frac{p\ \cos\theta}{4\pi\varepsilon_0 r^2}$$

where p is the dipole moment ($p = q\,\ell$) of the dipole, r is the distance from the dipole to the point at which we calculate V, and θ is as shown in the figure.

$$\frac{1}{4\pi\varepsilon_0}$$

is a constant equal to $9 \times 10^9 \mathrm{N} \times \mathrm{m}^2/\mathrm{C}^2$. The magnitude of the electric field intensity is

$$E = \frac{p}{4\pi\varepsilon_0 r^3}\sqrt{4 \cos^2\theta + \sin^2\theta}$$

If these are to be equal in magnitude,

$$\cos\theta = \frac{\sqrt{4 \cos^2\theta + \sin^2\theta}}{r}$$

or
$$r^2 = \frac{4 \cos^2\theta + \sin^2\theta}{\cos^2\theta} = 4 + \tan^2\theta$$

The minimum value of r^2 occurs when $\tan\theta = 0$. Hence, the minimum value of r for V and E to be equal in magnitude occurs for $r^2 = 4$; that is, $r = 2$ m, in MKS units.

If $r = \sqrt{5}$ m, then V and E are equal in magnitude when

$$\left(\sqrt{5}\right)^2 = \frac{4 \cos^2\theta + \sin^2\theta}{\cos^2\theta}$$

$$5 \cos^2\theta = 4\cos^2\theta + \sin^2\theta$$

$$\cos^2\theta = \sin^2\theta$$

Thus, $\theta = 45°$, $135°$, $225°$, or $315°$.

22.6 Electric Potential Energy

$$U = \frac{1}{4\pi\varepsilon_0} \frac{q_1 q_2}{r_{12}}$$

Here r_{12} is the distance between q_1 and q_2.

Problem Solving Example:

 Calculate the potential energy of the charge distribution shown in the diagram.

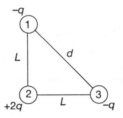

 The potential energy of any two charges q_i and q_j is

$$U_{ij} = k_E \frac{q_i q_j}{R_{ij}}$$

where R_{ij} is the distance between the two charges, and

$$k_E = 9 \times 10^9 \, \frac{\mathrm{N} \times \mathrm{m}^2}{\mathrm{C}^2} = \frac{1}{4\pi\varepsilon_0}$$

The potential energy of the charge distribution is the sum of the potential energies of every possible pair of charges within the distribution. Hence,

$$U = U_{12} + U_{13} + U_{23} = k_E \left[\frac{(-q)(2q)}{L} + \frac{(-q)(-q)}{d} + \frac{(+2q)(-q)}{L} \right]$$

Since $d^2 = L^2 + L^2$, we have $d = \sqrt{2L^2}$ and

$$U = k_E \left[\frac{-2q^2}{L} + \frac{q^2}{\sqrt{2L}^2} - \frac{2q^2}{L} \right] = \frac{k_e q^2}{L} \left[\frac{-2\sqrt{2} + 1 - 2\sqrt{2}}{\sqrt{2}} \right]$$

$$= -3.29 k_E \frac{q^2}{L} \text{ J}$$

If $q = e$ and $L = 1 \text{ A} = 10^{-8}$ m, then

$$U = -3.29 \cdot 9 \times 10^9 \, \text{n} \cdot \text{m}^2/\text{C}^2 \times \frac{\left(1.6 \times 10^{-19} \text{C}\right)^2}{10^{-8} \text{m}}$$

$$= -7.58 \times 10^{-20} \text{J}$$

The negative sign indicates that work would be required to disassemble the charge distribution (i.e., work must be done against attractive electrical forces).

22.7 Calculation of *E* from *V*

$$E_r = -\frac{dV}{dr}$$

where the electric field components for all points in space are:

$$E_x = \frac{-\partial V}{\partial x}, \quad E_y = \frac{-\partial V}{\partial y}, \quad \text{and } E_z = \frac{-\partial V}{\partial z}$$

and

$$r = \text{path to the *equipotential surface}$$

* Equipotential surface is one where the potential differences of a locus of points surrounding the surface are the same.

Quiz: Electric Potential (Voltage)

1. The diagram shows three point charges, each 1 meter apart. The best indication of the net force on the +1 charge is

    $$0 + 1$$

    $$+ 20 \qquad 0 + 2$$

    (A) to the right.

    (B) to the left.

    (C) straight down.

    (D) straight up.

    (E) down and to the right.

2. The potential energy change a unit charge experiences in moving from one point to another is called the

    (A) voltage.

    (B) current.

    (C) resistance.

    (D) field.

    (E) charge.

3. You have three identical neutral hollow metal spheres. You charge the first, then touch it to the second. Now touch the second to the third. Finally, touch the third to the first. What is the fraction of the original charge now on sphere one, two, and three?

    (A) $\dfrac{1}{4}, \dfrac{1}{2}, \dfrac{1}{4}$

    (B) $\dfrac{3}{8}, \dfrac{2}{8}, \dfrac{3}{8}$

    (C) $\dfrac{1}{3}, \dfrac{1}{3}, \dfrac{1}{3}$

    (D) $\dfrac{3}{8}, \dfrac{1}{4}, \dfrac{1}{4}$

    (E) $\dfrac{1}{4}, \dfrac{1}{4}, \dfrac{1}{8}$

QUESTIONS 4 and 5 refer to a parallel-plate capacitor with a plate separation of d and surface area A.

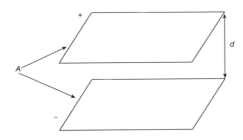

4. If the potential difference between the plates is V and the distance between the plates is d, the potential energy of the capacitor is

(A) $\dfrac{\varepsilon_0 A V^2}{2}$.

(D) $\dfrac{\varepsilon_0 A V}{2d}$.

(B) $\dfrac{\varepsilon_0 A V^2}{2d}$.

(E) $\dfrac{\varepsilon_0 A V}{d}$.

(C) $\dfrac{\varepsilon_0 A V^2}{d}$.

5. The plates are now separated an additional distance Δd. The work required to separate the plates is equal to

(A) $\left[\dfrac{\varepsilon_0 A V^2}{2}\right]\left[\dfrac{\Delta d}{d^2 + d\Delta d}\right]$.

(D) $\left[\dfrac{\varepsilon_0 A V^2}{2}\right]\left[\dfrac{\Delta d}{d + \Delta d}\right]$.

(B) $\left[\dfrac{\varepsilon_0 A V^2}{2}\right]\left[-\dfrac{\Delta d}{d^2 + d\Delta d}\right]$.

(E) $\left[\dfrac{\varepsilon_0 A V^2}{2}\right]\left[-\dfrac{\Delta d^2}{d^2 + d\Delta d}\right]$.

(C) $\left[\dfrac{\varepsilon_0 A V^2}{2}\right]\left[\dfrac{\Delta d}{d + \Delta d}\right]$.

6. Charges q_1 and q_2 are isolated and fixed in space. The amount of work necessary to bring q_3 from infinity to point C is

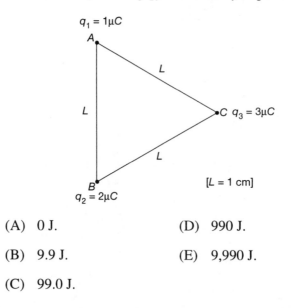

$q_1 = 1\mu C$

A

L

L

C $q_3 = 3\mu C$

L

B

$q_2 = 2\mu C$

[L = 1 cm]

(A) 0 J.

(D) 990 J.

(B) 9.9 J.

(E) 9,990 J.

(C) 99.0 J.

7. Figure out the total electric potential energy of a *single* spherical object of uniform charge density ρ, total charge Q, and radius R. Let $k = {}^1/_{(4\pi\varepsilon 0)}$ as usual.

(A) 0

(B) $\dfrac{kQ^2}{R}$

(C) $1/2\ kQ^2 / R$

(D) $3/5\ kQ^2 / R$

(E) $2/3\ kQ^2 / R$

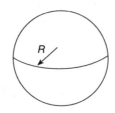

R

8. Determine the electric potential of the infinite sheet of charge shown below for $x > 0$. Let the charge density be σ and the x-direction be to the right.

(A) $\dfrac{-\sigma x}{\varepsilon_0}$

(B) $\dfrac{+\sigma x}{2\varepsilon_0}$

(C) $\dfrac{-\sigma x}{2\varepsilon_0}$

(D) $\dfrac{+\sigma x}{\varepsilon_0}$

(E) $\dfrac{-2\sigma}{\varepsilon_0}$

9. Find the potential energy per ion for an infinite one-dimensional ionic string of charges of magnitude e and alternating sign. Let the distance between the ions be s.

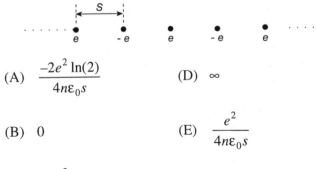

(A) $\dfrac{-2e^2 \ln(2)}{4n\varepsilon_0 s}$

(D) ∞

(B) 0

(E) $\dfrac{e^2}{4n\varepsilon_0 s}$

(C) $\dfrac{-e^2}{4n\varepsilon_0 s}$

10. Determine the electric potential of a circular annulus of inner radius a and outer radius b (in the yz plane) of charge density s = charge/area < 0 at a distance x along the x-axis.

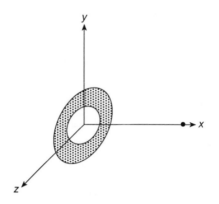

(A) $\Phi = \dfrac{\pi k \sigma [b^2 - a^2]}{x}$

(B) $\Phi = -2\pi k \sigma \left[\sqrt{b^2 + x^2} - \sqrt{a^2 + x^2} \right]$

(C) $\Phi = -2\pi k \sigma \left[\sqrt{b^2 + x^2} - x \right]$

(D) $\Phi = +2\pi k \sigma \left[\sqrt{b^2 + x^2} - \sqrt{a^2 + x^2} \right]$

(E) $\Phi = +2\pi k \sigma \left[\sqrt{b^2 + x^2} - x \right]$

ANSWER KEY

1.	(D)	6.	(B)
2.	(A)	7.	(D)
3.	(B)	8.	(C)
4.	(B)	9.	(A)
5.	(B)	10.	(D)

CHAPTER 23

Capacitors

23.1 Parallel-plate Capacitors

$$C = \frac{q}{V} = \frac{\varepsilon_0 A}{d} \quad \rightarrow \quad \text{units: } \frac{\text{coulomb}}{\text{volt}} = \text{farad}$$

C = capacitance

q = electric charge

V = electric potential

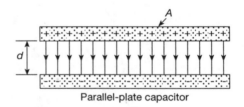

Parallel-plate capacitor

ε_0 = permittivity constant

A = cross-sectional area of plates

d = distance between plate surfaces

Problem Solving Example:

 The plates of a parallel plate capacitor are 5 mm apart and 2 m² in area. The plates are in vacuum. A potential difference of 10,000 volts is applied across the capacitor. Compute a) the capacitance, b) the charge on each plate, and c) the electric intensity.

 a)

$$C = \varepsilon_0 \frac{A}{d}$$

$$= 8.85 \times 10^{-12} \frac{F}{m} \times \frac{2\,\text{m}^2}{5 \times 10^{-3}\,\text{m}}$$

$$= 3.54 \times 10^{-9} \ \text{farad}$$

$$= 3.54 \times 10^{-3} \ \mu\text{F}$$

$$= 3{,}540 \ \text{pF}$$

b) The charge on the capacitor is

$$q = CV_{ab}$$

$$= (3.54 \times 10^{-9}\,\text{F})(1 \times 10^4\,\text{V})$$

$$= 3.54 \times 10^{-5}\,\text{C}$$

c) The electric intensity is

$$E = \frac{1}{\varepsilon_0} \sigma = \frac{1}{\varepsilon_0} \frac{q}{A}$$

$$= \left(36\pi \times 10^9 \, \frac{m}{F}\right) \times \left(1.77 \times 10^{-5} \, \frac{FV}{m^2}\right)$$

$$= 2 \times 10^6 \ \text{V/m}$$

The electric intensity may also be computed from the potential gradient.

$$E = \frac{V_{ab}}{d}$$

$$= \frac{10^4 \, \text{V}}{5 \times 10^{-3} \, \text{m}} = 2 \times 10^6 \, \text{V/m}$$

23.2 Cylindrical Capacitors

$$C = \frac{2\pi\varepsilon_0 \ell}{\ln\left(\dfrac{b}{a}\right)}$$

C = capacitance

ℓ = length of capacitor

ε_0 = permittivity constant

b = outer plate radius

a = inner plate radius

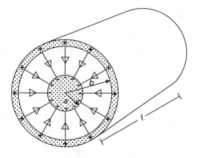

Figure 23.1
Cylindrical Capacitor

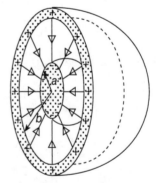

Figure 23.2
Spherical Capacitor

23.3 Spherical Capacitors

$$C = 4\pi\varepsilon_0 \frac{ab}{b-a}$$

C = capacitance

ε_0 = permittivity constant

a = inner plate radius

b = outer plate radius

23.4 Capacitors in Series

$$\frac{1}{C} = \sum_n \frac{1}{C_n}$$

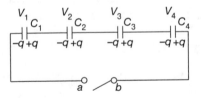

Example: Circuit with capacitors in series

After the switch is connected, the total capacitance may be found.

$$\frac{1}{C_{total}} = \frac{1}{C_1} + \frac{1}{C_2} + \frac{1}{C_3} + \frac{1}{C_4}$$

23.5 Capacitors in Parallel

$$C = \sum_n C_n$$

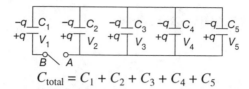

$$C_{\text{total}} = C_1 + C_2 + C_3 + C_4 + C_5$$

Problem Solving Examples:

 In the figure below, $C_1 = 6 \,\mu F$, $C_2 = 3 \,\mu F$ and $V_{ab} = 18$ V. Find the charge on each capacitor. What is the value of the equivalent capacitance (see figure)?

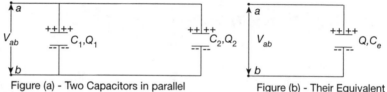

Figure (a) - Two Capacitors in parallel   Figure (b) - Their Equivalent

A The charge Q on a capacitor having capacitance C is $Q = VC$, where V is the potential difference across the capacitor. Then,

$$Q_1 = V_{ab}C_1 = (18 \ V) \times (6 \times 10^{-6} \ F) = 108 \times 10^{-6} \ C = 1.08 \times 10^{-4} C$$

$$Q_2 = V_{ab}C_2 = (18 \ V) \times (3 \times 10^{-6} \ F) = 54 \times 10^{-6} \ C = 5.4 \times 10^{-5} C$$

The equivalent capacitance must carry the same total charge as the original system, since charge is conserved (none leaks out of the system). Hence,

$$Q_{\text{net}} = Q_1 + Q_2 = 162 \times 10^{-6} \ C = 1.62 \times 10^{-4} C$$

Then, the equivalent capacitance is

$$C_e = \frac{Q_{\text{net}}}{V_{ab}} = \frac{162 \times 10^{-6} C}{18 \ V} = 9 \times 10^{-6} F$$

How should five capacitors, each of capacitance 1 μF, be connected so as to produce a total capacitance of $^3/_7$ μF?

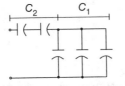

Series connection of C_1 and C_2

If all capacitors are joined in parallel, the resultant capacitance is 5 μF. (For the resultant capacitance of a set of capacitors connected in parallel equals the sum of the individual capacitances.) If the capacitors are connected in series, the resultant capacitance is $^1/_5$ μF (for the resultant capacitance of a set of capacitors connected in series equals the reciprocal of the sum of the reciprocals of the individual capacitances). The connection is thus more complicated.

Suppose that n capacitors are connected in parallel and $5 - n$ in series. The resultant capacitances are thus such that

$$C_1 = \underbrace{(1+1+ \ \ldots)}_{n \text{ times}} \mu F = n \mu F$$

and

$$\frac{1}{C_2} = \underbrace{\left(\frac{1}{1}+\frac{1}{1}+ \ \ldots \right)}_{(5-n) \text{ times}} \mu F = (5-n) \mu F^{-1}$$

or

$$C_2 = \frac{1}{5-n} \mu F$$

If C_1 and C_2 are connected in parallel, then

$$C_r = C_1 + C_2 = \left(n + \frac{1}{5-n} \right) \mu F = \frac{3}{7} \ \mu F$$

$$5n - n^2 + 1 = \frac{15 - 3n}{7}$$

$$35n - 7n^2 + 1 = 15 - 3n$$

$$7n^2 - 38n + 8 = 0$$

By the quadratic formula, $n = \dfrac{-(-38) \pm \sqrt{(-38)^2 - 4(7)(8)}}{2(7)}$. There is thus no integral solution for n.

But if C_1 and C_2 are connected in series (see figure on previous page), then

$$\frac{1}{C_r} = \frac{1}{C_1} + \frac{1}{C_2} = \left(\frac{1}{n} + 5 - n\right) \mu F^{-1} = \frac{7}{3} \mu F^{-1}.$$

$$\therefore \ 3 + 15n - 3n^2 = 7n \ \text{ or } \ 3n^2 - 8n - 3 = 0.$$

$$\therefore \ (3n + 1)(n - 3) = 0.$$

This has an integral solution for n, $n = 3$. Thus, the required capacitance is given if three capacitors are connected in parallel, and the combination is connected in series with the other two.

23.6 Energy Stored in a Capacitor

$$U_T = \frac{1}{2}CV^2$$

U_T = total energy

C = capacitance

V = electric potential difference

Problem Solving Example:

Q A plane parallel plate capacitor consisting of two metal circular plates 5 cm in radius separated 1 mm in air is charged to 300 stat-volts, whereupon it is connected in parallel to another simi-

larly charged capacitor (positive terminals connected together and negative terminals connected) (see Figure (a)). How much energy would be
released if the combinations were discharged by a short circuit?

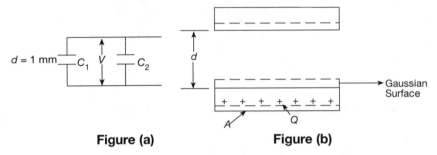

| Figure (a) | Figure (b) |

 For a plane parallel plate capacitor:

$$C = \frac{kA}{4\pi d}$$

This result can be obtained as follows. According to the definition
of capacitance C,

$$C = \frac{Q}{V}$$

where Q is the total charge on one plate, and V is the potential difference between the plates. According to Gauss' law, if the Gaussian Surface is constructed as shown in Figure (b), then

$$\frac{k}{4\pi} \phi_E = \frac{k}{4\pi} EA = Q \tag{1}$$

in the CGS system.

This relation holds because the electric field E is a constant in the
parallel plate capacitor. Q is the charge enclosed by the Gaussian Surface. It is also the total charge on either plate, due to the construction of
the surface. Also, for a parallel plate capacitor

$$V = Ed \tag{2}$$

Therefore, combining (1) and (2), we get the result for the capacitance of the parallel plate capacitor.

$$\therefore \ C_1 = \frac{K\pi r^2}{4\pi d} = \frac{5^2}{4 \times 1 \ mm \times 1 \ cm/10 \ mm}$$

$$= \frac{25}{.4} = 62.5 \ \ stat\text{--}farads$$

Recalling that the energy stored in a capacitor is

$$U = \frac{1}{2}QV = \frac{1}{2}CV^2 = \frac{1}{2}\frac{Q^2}{C}$$

and choosing the second form because C and V are known

$$W_1 = \frac{1}{2} \ C_1 V_1^2 = \frac{1}{2} \ (62.5)(300)^2 = 31.25 \ (90,000)$$

$$= 2,820,000 \ \ ergs$$

But the total energy stored in the two capacitors is

$$W_1 + W_2 = 2W_1$$

$$\therefore \ W = 2(2,820,000) = 5,640,000 \ \ ergs$$

$$= .564 \ \ joules$$

23.7 Energy Density in an Electric Field

$$u = \frac{1}{2}\varepsilon_0 E^2$$

u = energy density (stored energy per unit volume)

ε_0 = permittivity constant

E = electric field

23.8 Parallel-plate Capacitor with Dielectric

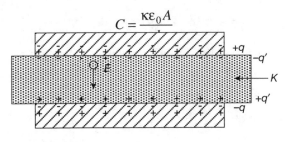

$$C = \frac{\kappa \varepsilon_0 A}{d}$$

C = capacitance

κ = dielectric constant (specific to the material)

ε_0 = permittivity constant

A = area of plates

d = distance between plates

Problem Solving Example:

  A parallel plate capacitor of 2 meter^2 in area and with charge $q = 3.54 \times 10^{-5}$ coulomb is insulated while a sheet of dielectric 5 mm thick, of dielectric constant 5, is inserted between the plates. Compute a) the electric intensity in the dielectric, b) the potential difference across the capacitor, and c) its capacitance.

a) The insertion of a dielectric between the capacitor plates alters the electric intensity because of the reversed field set up by the induced charges on the dielectric.

The electric intensity is

$$E = \frac{\sigma}{\kappa \varepsilon_0} = \frac{1}{\kappa \varepsilon_0} \frac{q}{A}$$

$$= \frac{1}{5(8.85 \times 10^{-12}\,\text{F/m})} \frac{3.54 \times 10^{-5}\text{C}}{2\,\text{m}^2}$$

$$= 4 \times 10^5 \quad \text{volts/meter}$$

b) The potential difference across the capacitor is reduced to

$$V_{ab} = Ed$$
$$= (4 \times 10^5\,\text{V/m})(5 \times 10^{-3}\,\text{m})$$
$$= 2,000 \quad \text{volts}$$

c) The capacitance is increased to

$$C = \frac{q}{V_{ab}}$$

$$= \frac{3.54 \times 10^{-5}}{2,000}$$

$$= 1.77 \times 10^{-8}\,\text{F}$$

$$= 17,700\,\text{pF}$$

The capacitance may also be computed from

$$C = \varepsilon \frac{A}{d} = \kappa \varepsilon_0 \frac{A}{d}$$

$$= 5(8.85 \times 10^{-12}\,\frac{\text{F}}{\text{m}}) \frac{2\,\text{m}^2}{5 \times 10^{-3}\,\text{m}}$$

$$= 17,700\,\text{pF}$$

23.9 Cylindrical Capacitor with Dielectric

$$C = \frac{k\varepsilon_0 2\pi\ell}{\ln\left(\dfrac{b}{a}\right)}$$

C = capacitance

κ = dielectric constant

ε_0 = permittivity constant

ℓ = length of plates

b = radius of outer plate

a = radius of inner plate

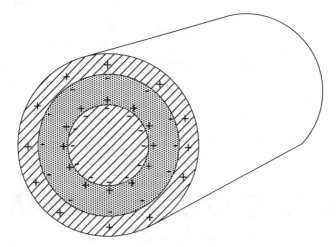

Figure 23.3 Cylindrical Plates with a Dielectric

23.10 Energy Stored in a Capacitor with a Dielectric

$$U = \frac{1}{2}\kappa CV^2$$

where κ is the dielectric constant.

23.11 Energy Density in an Electric Field with a Dielectric

$$u = \frac{1}{2}\kappa\varepsilon_0 E^2$$

where κ is the dielectric constant.

23.12 Gauss' Law with a Dielectric

$$q = \varepsilon_0 \oint (\kappa\overline{E}) \bullet d\overline{S}$$

q = total charge within the surface

ε_0 = permittivity constant

κ = dielectric constant

E = electric field

S = surface area

Problem Solving Example:

Q The space between the plates of a parallel-plate capacitor is filled with dielectric of coefficient 2.5 and strength $5 \times 10^6 V \times m^{-1}$. The plates are 2 mm apart. What is the maximum voltage that can be applied between the plates? What area of plates will give a capacitance of $10^{-3}\mu F$? At the maximum voltage what are the free and bound charges per unit area of the plate and dielectric surface?

A The dielectric strength of the dielectric is the largest electric field which it can withstand before becoming a conductor. We must relate the voltage between the capacitor plates to E_{max} (dielectric strength). This can be done by realizing that voltage differences are defined by

$$V_b - V_a = -\int_a^b \overline{E} \cdot d\overline{\ell} \tag{1}$$

where V_b and V_a refer to the potentials at $z = b$ and $z = a$, respectively (see figure), and $\overline{E}$ is the electric field. The integral in (1) is a line integral, and it is to be evaluated over an arbitrary path connecting a and b, $d\overline{\ell}$ being a small element of path length. Looking at the figure, we see that $\overline{E}$ is composed of contributions from two sources—the conducting plates ($\overline{E}_c$) and the dielectric material ($\overline{E}_D$). Note that the latter is in opposition to $\overline{E}_c$. Since $\overline{E}$ is uniform in direction and magnitude for a parallel-plate capacitor, (1) becomes

$$V_b - V_a = -\overline{E} \cdot \int_a^b d\overline{\ell} \tag{2}$$

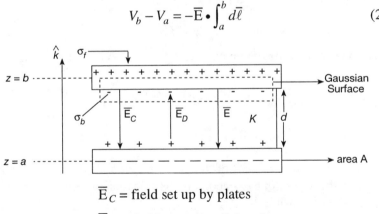

$\overline{E}_C$ = field set up by plates

$\overline{E}_D$ = field set up by dielectric

$\overline{E}$ = net field = $E_D + E_c$

The easiest path to evaluate this line integral along is a straight line path from $z = a$ to $z = b$. In this case,

$$d\bar{\ell} = d\bar{\ell} \; \hat{k}$$

Since

$$\bar{E} = -E \; \hat{k}$$

(2) becomes

$$V_b - V_a = -(-E\hat{k}) \bullet \int_a^b (d\ell \; \hat{k})$$
$$V_b - V_a = E(b-a)$$

But $b - a = d$ (see figure on previous page) and we define

$$V_b - V_a = \Delta V$$

as the potential difference between the plates. Therefore,

$$\Delta V = Ed \tag{3}$$

The maximum voltage that can be applied between the plates is

$$\Delta V_{max} = E_{max}d = \left(5 \times 10^6 \, \frac{V}{m}\right)\left(2 \times 10^{-3} \, m\right)$$
$$\Delta V_{max} = 10^4 \, V$$

To answer the second part of the problem, we must relate the capacitance of the capacitor to its geometry. We do this by beginning with the definition of capacitance (C)

$$C = \frac{Q}{\Delta V} \tag{4}$$

where Q is the charge on one capacitor plate. If the free charge density on the plate of cross-sectional area A is σ_f (see figure), then

$$Q = \sigma_f A \tag{5}$$

Using (5) and (3) in (4),

$$C = \frac{\sigma_f A}{Ed} \qquad (6)$$

Since we don't know E in this portion of the problem, we must eliminate it from (6). This may be done by using Gauss' Law for a dielectric

$$\oint_s \overline{E} \cdot d\overline{S} = \frac{Q_{net}}{\varepsilon} \qquad (7)$$

where ε is a constant, and Q_{net} is the total free charge enclosed by the surface s over which the integral is evaluated. Evaluating (7) over the Gaussian surface indicated in the figure,

$$\oint_s \overline{E} \cdot d\overline{S} = \int_{s_T} \overline{E}(s_T) \cdot d\overline{S} + \int_{s_B} \overline{E}(s_B) \cdot d\overline{S} = \frac{Q_{net}}{\varepsilon} \qquad (8)$$

where s_T and s_B are the top and bottom of the rectangular surface. Now, $d\overline{S}$ is a vector element of surface area. For s_T, $d\overline{S} = ds\,\hat{k}$ and for s_B, $d\overline{S} = -ds\,\hat{k}$. Then, using the fact that $\overline{E} = -E\hat{k}$ everywhere, (8) becomes

$$\oint_s \overline{E} \cdot d\overline{S} = \int_{s_T} \left(-E(s_T)\hat{k}\right) \cdot \left(ds\,\hat{k}\right) + \int_{s_B} \left(-E(s_B)\hat{k}\right) \cdot \left(-ds\,\hat{k}\right) = \frac{Q_{net}}{\varepsilon}$$

$$\oint_s \overline{E} \cdot d\overline{S} = -\int_{s_T} E(s_T)ds + \int_{s_B} E(s_B)ds = \frac{Q_{net}}{\varepsilon}$$

Now, $E(s_T)$ is the value of E at s_T, which lies in the top conducting plate. By definition of a conductor, then, $E(s_T) = 0$. Therefore,

$$\oint_s \overline{E} \cdot d\overline{S} = \int_{s_B} E(s_B)ds = \frac{Q_{net}}{\varepsilon} \qquad (9)$$

But $E(s_B) = E$, the electric field between the capacitor plates. This is constant, and (9) becomes

$$E \int_{s_B} ds = EA = \frac{Q_{net}}{\varepsilon}$$

since the area of s_B is A. Hence,

$$E = \frac{Q_{net}}{\varepsilon A}.$$

Now, Q_{net} includes the free charge found within the Gaussian surface. Therefore (see figure on page 397),

and
$$E = \frac{\sigma_f}{\varepsilon} \tag{10}$$

Using (10) in (6),

$$C = \frac{\sigma_f A}{\dfrac{\sigma_f}{\varepsilon} d} = \frac{\varepsilon A}{d}$$

But κ, the dielectric constant, is defined by

$$\kappa = \frac{\varepsilon}{\varepsilon_0}$$

Hence, $\varepsilon = \varepsilon_0 \kappa$ and

$$C = \frac{\kappa \varepsilon_0 A}{d} \tag{11}$$

Solving for A,

$$A = \frac{Cd}{\kappa \varepsilon_0} \tag{12}$$

Substituting the given data in (12),

$$A = \frac{(10^{-3}\,\mu F)(2\times10^{-3}\ m)}{(2.5)(8.85\times10^{-12}\,c^2/N\times m^2)}$$

$$A = .09\ m^2$$

The free charge density can be found using (10)

$$\sigma_f = \varepsilon E = \varepsilon_0 \kappa E$$

or

$$\sigma_f = (5\times10^6\ \text{V/m})(2.5)(8.85\times10^{-12}\,C^2/N\times m^2)$$

$$\sigma_f = 110.6\times10^{-6}\,C/m^2$$

$$\sigma_f = 1.106\times10^{-4}\,C/m^2$$

From the figure on page 397, note that the net field $\overline{E}$ is the superposition of the field due to the bound charges of the dielectric ($\overline{E}_D$) and the field due to the free charge on the capacitor plates ($\overline{E}_c$). Then

$$\overline{E} = \overline{E}_C + \overline{E}_D \tag{13}$$

But, by (10) and the figure,

$$\overline{E} = -\frac{\sigma_f}{\varepsilon}\hat{k}$$

Since the bound charge accumulates only on the surface of the dielectric, we may consider the dielectric slab to be equivalent to a parallel plate capacitor with surface density σ_b and an air dielectric.

$$\overline{E}_D = -\frac{\sigma_b}{\varepsilon_0}\hat{k}$$

Similarly,

$$\overline{E}_C = -\frac{\sigma_f}{\varepsilon_0}\hat{k}$$

Hence, from (13)

$$-\frac{\sigma_f}{\varepsilon}\hat{k} = -\frac{\sigma_f}{\varepsilon_0}\hat{k} + \frac{\sigma_b}{\varepsilon_0}\hat{k}$$

or

$$-\sigma_f = -\sigma_f\frac{\varepsilon}{\varepsilon_0} + \sigma_b\frac{\varepsilon}{\varepsilon_0}$$

Since

$$\kappa = \frac{\varepsilon}{\varepsilon_0}$$

$$\sigma_b = \sigma_f\frac{(K-1)}{K} = \frac{1.5}{2.5}\sigma_f = \frac{3}{5}\sigma_f$$

Noting that

$$\sigma_f = 110.6 \times 10^{-6}\, \text{C/m}^2$$

$$\sigma_b = \frac{331.8}{5} \times 10^{-6}\, \text{C/m}^2$$

$$\sigma_b = 6.64 \times 10^{-5}\, \text{C/m}^2$$

Quiz: Capacitors

QUESTIONS 1–3 refer to three 1 microfarad capacitors and the following possible results.

    I. 0.33 microfarad

    II. 1.0 microfarad

    III. 0.55 microfarad

    IV. 3.0 microfarads

    V. 6.0 microfarads

1.   If the three capacitors are connected in parallel, their total capacitance will be

    (A) I.   (D) IV.

    (B) II.   (E) V.

    (C) III.

2.   If the three capacitors are connected in series, their capacitance will be

    (A) I.   (D) IV.

    (B) II.   (E) V.

    (C) III.

3.   If the dielectric ($\kappa = 1.0$) in each capacitor is replaced with a dielectric ($\kappa = 3.0$), the capacitance of each capacitor will be

(A) I.   (D) IV.

(B) II.   (E) V.

(C) III.

4.   Two 4 μF parallel plate capacitors are connected in parallel across a 12 V battery. What is the total amount of charge stored in the two capacitors?

(A) 48 μC   (D) 24 μC

(B) 96 μC   (E) 0.6 μC

(C) 0.3 μC

5.   A parallel plate capacitor is connected to a fixed voltage. If a dielectric material is introduced between the plates, which one of the following will result?

(A) Charge decreases and capacitance increases.

(B) Charge and capacitance decrease.

(C) Charge increases and capacitance decreases.

(D) Charge and capacitance increase.

(E) Charge and capacitance are not changed.

6.   What voltage is necessary to produce a 300 μC charge on a 4 μF capacitor?

(A) 0.013 V   (D) 300 V

(B) 4 V   (E) 1,200 V

(C) 75 V

7. If the distance between the plates of a parallel plate capacitor is doubled, what must you do to the area of the plates to return to the original capacitance?

   (A) Cut the plate area in half.

   (B) Cut the plate area by one-fourth.

   (C) Double the plate area.

   (D) Make no change because the change in distance between the plates does not change the capacitance.

   (E) Increase the area by four times.

8. Two capacitors having capacitances of 3 and 6 farads respectively are connected in parallel. The total capacitance in farads is

   (A) 9.   (D) .5.

   (B) 2.   (E) 45.

   (C) 18.

9. Two capacitors arranged in a parallel connection have equivalent capacitance that can be calculated by the formula $C_{equiv.} =$

   (A) $\dfrac{1}{C_1} + \dfrac{1}{C_2}$.   (D) $C_1 \times C_2$.

   (B) $C_1 + C_2$.   (E) $\dfrac{C_1}{C_2}$.

   (C) $(C_1 + C_2)/2$.

10. Three capacitors of 2 farads each are connected in parallel. The equivalent capacitance is

(A) 2.0 farads.   (D) 6.0 farads.

(B) 1.5 farads.   (E) 8.0 farads.

(C) 0.667 farad.

ANSWER KEY

1.	(D)		6.	(C)
2.	(A)		7.	(C)
3.	(D)		8.	(A)
4.	(B)		9.	(B)
5.	(B)		10.	(D)

CHAPTER 24

Current and Resistance

24.1 Electric Current

$$i = \frac{dq}{dt} \quad \rightarrow \quad \text{units: amperes}$$

i = electric current

dq = differential charge

dt = differential time

24.2 Current Density and Current

$$j = \frac{i}{A} \quad \rightarrow \quad \text{units:} \ \frac{\text{amperes}}{\text{meters}^2}$$

j = current density

i = electric current

A = cross-sectional area

$$i = \int \bar{j} \cdot d\overline{S}$$

where $\overline{S}$ is surface area.

24.3 Mean Drift Speed

$$\overline{v}_D = \frac{\bar{j}}{ne}$$

v_D = mean drift speed

j = current density

n = number of atoms per unit volume

e = electron charge

Problem Solving Example:

A car battery supplies a current I of 50 amp to the starter motor. How much charge passes through the starter in $^1/_2$ min?

Current (I) is defined as the net amount of charge, Q, passing a point per unit time. Therefore,

$$Q = It = (50 \ \text{amp}) \ (30 \ \text{sec}) = 1,500 \ \text{coul.}$$

24.4 Resistance

$$R = \frac{V}{I} \quad \text{units} \ \rightarrow \ \text{ohms} \ (\Omega)$$

R = resistance

V = electric potential difference (voltage)

I = current

Problem Solving Examples:

 What is the resistance of an electric toaster if it takes a current of 5 amperes when connected to a 120-volt circuit?

 This is an application of Ohm's law. Since we wish to find the resistance, we use

$$R = \frac{V}{I}$$

$$V = 120 \text{volts}, \quad I = 5 \text{amperes}$$

Therefore,

$$R = \frac{120 \text{ volts}}{5 \text{ amperes}} = 24 \text{ ohms}$$

 The voltage across the terminals of a resistor is 6.0 volts and an ammeter connected as in the diagram reads 1.5 amp. a) What is the resistance of the resistor? b) What would the current be if the potential difference were raised to 8.0 volts?

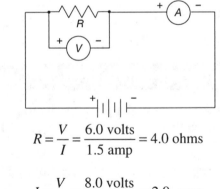

a) $$R = \frac{V}{I} = \frac{6.0 \text{ volts}}{1.5 \text{ amp}} = 4.0 \text{ ohms}$$

b) $$I = \frac{V}{R} = \frac{8.0 \text{ volts}}{4.0 \text{ ohms}} = 2.0 \text{ amp}$$

In part a) of this solution, we have used merely the definition of resistance. But in part b) we have used Ohm's law, that is, the fact that R is constant.

24.5 Resistivity

$$\rho = \frac{E}{j} \quad \rightarrow \text{units: ohm - meters } (\Omega\text{m})$$

ρ = Resistivity

E = Electric Field

j = Current Density

For a cylindrical conductor,

$$R = \rho \frac{L}{A}$$

R = resistance

ρ = resistivity

L = length

A = cross-sectional area

Problem Solving Example:

 What is the resistance of a piece of nichrome wire 225 centimeters long with a cross-sectional area of 0.015 square centimeter?

To solve this problem we use the relationship

$$R = \rho \frac{L}{A}$$

where

R = resistance

ρ = resistivity

L = wire length

A = cross-sectional area

This basic relationship tells us that resistance is directly proportional to resistivity and length and inversely proportional to cross-sectional area. In the case of a wire, this means that the resistance depends on the nature of the substance (which appears in the equation as the resistivity) and that the resistance increases as the wire gets longer and decreases as the wire gets thicker.

The resistivity (ρ) for nichrome is 100×10^{-6} ohm-centimeter. The length is 225 centimeters, and the area is 0.015 square centimeter. Then

$$R = \frac{10^{-4} \text{ ohm-cm} \times 225 \text{ cm}}{0.015 \text{ cm}^2} = 1.5 \text{ ohms}$$

24.6 Temperature and Resistivity

$$\rho = \rho_0\left[1 + \alpha(T - T_0)\right]$$

ρ = final resistivity

ρ_0 = initial resistivity

α = coefficient of resistivity

T = final temperature

T_0 = initial temperature

Problem Solving Example:

 The resistance of a copper wire 2,500 cm long and 0.090 cm in diameter is 0.67 ohm at 20°C. What is the resistivity of copper at this temperature?

 The resistivity of a conductor is directly proportional to the cross-sectional area A and its resistance, and inversely proportional to its length ℓ. Therefore, knowing the resistance, we have

$$\rho = R \frac{A}{\ell} = \frac{0.67 \, \text{ohm}}{2,500 \, \text{cm}} \pi (0.090 \, \text{cm})^2$$

$$= 6.8198 \times 10^{-6} \, \Omega\text{cm}$$

Resistivity is a characteristic of a material as a whole, rather than of a particular piece of it. For a given temperature, it is constant for the particular medium, as opposed to resistance which depends on the dimensions of the piece.

24.7 Calculation of a Metal's Resistivity

$$\rho = \frac{m}{ne^2 t^*}$$

ρ = resistivity

m = mass

n = number of atoms per unit volume

e = electron charge

t^* = mean time between electron collisions

Problem Solving Example:

 A silver wire has a resistance of 1.25 ohms at 0°C and a temperature coefficient of resistance of 0.00375C°. To what temperature must the wire be raised to double the resistance?

 The change in resistance $R_t - R_0$ is directly proportional to the change in temperature and the original resistance R_0. It can be shown from previous equations that:

$$\Delta R = \alpha R_0 \Delta T$$

Substituting values, we have

$$\Delta T = \frac{\Delta R}{\alpha R_0} = \frac{R_t - R_0}{\alpha R_0}$$

since

$$R_t = 2R_0 = 2(1.25) = 2.50$$

$$T - 0°C = \frac{(2.50 - 1.25) \text{ ohms}}{0.00375C° \times 1.25 \text{ ohms}}$$

$$T = 267°C$$

24.8 Power

$$P = VI \quad \rightarrow \quad \text{units: watts (w)}$$

P = power

I = current

V = electric potential difference (Voltage)

24.9 Joule's Law

$$P = I^2R \quad \text{or} \quad \frac{V^2}{R} \quad \rightarrow \quad \text{units: watts}$$

P = power

I = current

R = resistance

V = electric potential difference (voltage)

Problem Solving Examples:

 An automobile battery produces a potential difference (or "voltage") of 12 volts between its terminals. (It really consists of six 2-volt batteries following one after the other.) A headlight bulb is to be connected directly across the terminals of the battery and dissipate 40 watts of joule heat. What current will it draw and what must its resistance be?

A To find the current, we use the formula $P = IV$, where P is the power dissipated (40 watts), I is the current, and V is the voltage.

Therefore,

$$I = \frac{P}{V} = \frac{40 \text{ watts}}{12 \text{ volts}} = \frac{40 \text{ joules/sec}}{12 \text{ joules/coulomb}} = 3.33 \text{ coulombs/sec}$$

$$= 3.33 \text{ amps}$$

The bulb draws 3.33 amps from the battery.

From Ohm's law

$$I = \frac{V}{R}$$

$$R = \frac{V}{I}$$

$$R = \frac{12 \, \text{volts}}{3.33 \, \text{amps}}$$

$$R = 3.6 \, \text{ohms}$$

The resistance of the bulb must be 3.6 ohms.

We may also compute the resistance using the formula

$$P = \frac{V^2}{R}. \text{ Therefore}$$

$$R = \frac{V^2}{P} = \frac{(12)^2}{40} = 3.6 \ \text{ohms}$$

Q An electric kettle contains 2 liters of water which it heats from 20°C to boiling point in 5 min. The supply voltage is 200 V and a kWh (kilowatt-hour) unit costs 2 cents. Calculate (a) the power consumed (assume that heat losses are negligible), (b) the cost of using the kettle under these conditions six times, (c) the resistance of the heating element, and (d) the current in the element.

A The heat gained by the water in being raised to the boiling point is given by the expression $H = mc(t_2 - t_1)$, where c is the specific heat of water (the amount of heat required to raise the temperature of the substance 1°C), m is the mass of the water being heated, and $(t_2 - t_1)$ is the temperature difference before and after heating.

(a)

$$H = (2 \times 10^3 \, \text{cm}^3)(1 \text{g}/\text{c}^3)(4.18 \ \frac{\text{J}}{\text{g-}^\circ\text{C}}(100 - 20)^\circ\text{C}^\circ = 6.69 \times 10^5 \, \text{J}$$

Since heat losses are neglected, the conservation of energy requires that the heat energy generated be equal to the electrical energy consumed by the kettle.

Thus, the electric energy $E = 6.69 \times 10^5$ J,

The power is the energy consumed per second, which is thus

$$P = \frac{H}{t} = \frac{6.69 \times 10^5 \text{J}}{5 \min(60 \text{s/min})} = 2,230 \text{J/s} = 2.23 \times 10^3 \text{W} = 2.23 \ \text{kW}$$

(for 1W=1 J/s)

(b) The kettle uses 2.23 kW for 5 min each time the water is boiled. When it is used six times, 2.23 kW is used for 30 min = $^1/_2$ hr. The cost is thus

$$(2.23 \ \text{kW})(0.5 \, \text{hr})\left(\frac{2\cent}{\text{kW hr}} \right) = 2.23\cent$$

(c) The power P consumed is 2.23 kW and the supply voltage V is 200 V. But $P = V^2/R$, where R is the resistance of the kettle's heating element.

$$R = \frac{V^2}{P} = \frac{200^2 \ \text{V}^2}{2.23 \times 10^3 \ \text{W}} = 17.9 \ \Omega$$

(d) But one may also write the power as $P = IV$, where I is the current through the heating element.

$$I = \frac{P}{V} = \frac{2.23 \times 10^3 \text{W}}{200 \text{V}} = 11.2 \ \text{A}$$

CHAPTER 25

Circuits

25.1 Electromotive Force, EMF(e)

$$\varepsilon = \frac{dW}{dq}$$

ε = Electromotive Force

W = Work Done on Charge

q = Electric Charge

25.2 Current in a Simple Circuit

$$i = \frac{\varepsilon}{R}$$

i = Current

ε = Electromotive Force

R = Resistance

25.3 The Loop Theorem

$$\Delta V_1 + \Delta V_2 + \Delta V_3 \dots = 0$$

For a complete circuit loop

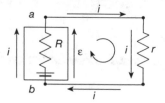

Figure 25.1 Simple Circuit with Resistor

Example:

$$V_b = -iR - ir + \varepsilon + V_b$$

Then

$$\varepsilon - iR - ir = 0$$

Note: If a resistor is traversed in the direction of the current, the voltage change is represented as a voltage drop, $-iR$. A change in voltage while traversing the EMF (or battery) in the direction of the EMF is a voltage rise $+\varepsilon$.

Problem Solving Examples:

Q In the figure $\varepsilon_1 = 12$ volts, $r_1 = 0.2$ ohm; $\varepsilon_2 = 6$ volts, $r_2 = 0.1$ ohm; $R_3 = 1.4$ ohms; and $R_4 = 2.3$ ohms. Compute a) the current in the circuit, in magnitude and direction, and b) the potential difference V_{ac}.

A a) From conservation of energy, we know that the sum of the changes in potential (voltage changes) around any closed loop must equal zero. Therefore, the current (which is conventionally taken as the flow of positive charge) is in a clockwise direction because $\varepsilon_1 > \varepsilon_2$; but let us choose it as going counterclockwise to show that it doesn't

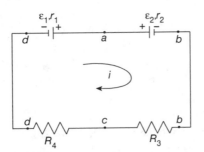

make any difference. Note that each battery has an internal resistance. Starting at point *a* and traversing counterclockwise yields the following equation:

$$-\varepsilon_1 - ir_1 - iR_4 - iR_3 - ir_2 + \varepsilon_2 = 0$$

$$i(r_1 + r_2 + R_3 + R_4) = \varepsilon_2 - \varepsilon_1$$

$$i = \frac{\varepsilon_2 - \varepsilon_1}{(r_1 + r_2 + R_3 + R_4)} = \frac{-6}{4}$$

$$= -1.5 \text{ amps}$$

The negative value for the current merely means that we chose the wrong direction (i.e., the current flows clockwise).

b)   We may use either a clockwise or counterclockwise path from point *a* to point *c* to find V_{ac}. Since we are only concerned with differences in potential, let us assume a zero potential of A and then traverse the loop clockwise from *a* to *c*. Taking into account the clockwise flow of current this yields,

$$-\varepsilon_2 - ir_2 - iR_3 = -6 - (1.5)(.1) - (1.5)(1.4)$$

$$= -8.25 \text{ volts}$$

This means that the potential is 8.25 volts lower at point *c* than at point *a*. If we go from point *a* to point *c* in a counterclockwise direction, we must remember to use the negative value for the current. This path yields:

$$-\varepsilon_1 - ir_1 - iR_4 = -12 - (-1.5)(.2) - (-1.5)(2.3)$$
$$= -12 + .3 + 3.45 = -8.25 \text{ volts}$$

Again, we see that the potential at point c is 8.25 volts lower than that at point a.

 Two devices, whose resistances are 2.8 and 3.5Ω, respectively, are connected in series to a 12-V battery. Compute the current in either device and the potential applied to each.

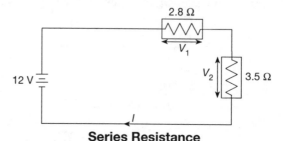

Series Resistance

 Resistances in series add. The equivalent resistance of the two devices is given by

$$R = 2.8\Omega + 3.5\Omega = 6.3\Omega$$

The current supplied by the battery can be computed using Ohm's law,

$$i = \frac{V}{R} = \frac{12V}{6.3\Omega} = 1.9\text{A}$$

This current i flows through both devices since they are in series. Ohm's law may be applied to calculate the potential applied to each device (see figure):

$$V_1 = iR_1 = (1.9\text{A})(2.8\Omega) = 5.3 \text{ V}$$
$$V_2 = iR_2 = (1.9\text{A})(3.5\Omega) = 6.7 \text{ V}$$

Note that the sum of these potentials is equal to the battery potential, 12 V.

25.4 Circuit with Several Loops

$$\sum_n i_n = 0$$

Example:

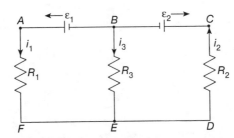

Figure 25.2 Multilogs Circuit $i_1 + i_2 + i_3 = 0$

Problem Solving Example:

 Suppose that three devices are connected in parallel to a 12-V battery. Let the resistances of the devices be $R_1 = 2\Omega$, $R_2 = 3\Omega$, and $R_3 = 4\Omega$. What current is supplied by the battery, and what is the current in each device?

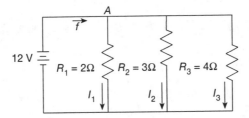

A Since the resistors are in parallel, we have for the equivalent resistance R,

$$\frac{1}{R} = \frac{1}{R_1} + \frac{1}{R_2} + \frac{1}{R_3} = \frac{1}{2\Omega} + \frac{1}{3\Omega} + \frac{1}{4\Omega} = \frac{13}{12}\Omega^{-1}$$

The equivalent resistance, therefore, is

$$R = \frac{12}{13}\Omega = 0.92\Omega.$$

The voltage across each device is 12 volts since they are in parallel to the battery. Therefore, using Ohm's law and the figure on the previous page,

$$I_1 = \frac{V}{R_1} = \frac{12\ V}{2\Omega} = 6\ \text{amp}$$

$$I_2 = \frac{V}{R_2} = \frac{12\ V}{3\Omega} = 4\ \text{amp}$$

$$I_3 = \frac{V}{R_3} = \frac{12\ V}{4\Omega} = 3\ \text{amp}$$

The current supplied by the battery is found by applying Kirchoff's node equation at point A. Hence,

$$I = I_1 + I_2 + I_3 = (6 + 4 + 3)\ \text{amps}$$
$$I = 13\ \text{amps}$$

or the current supplied by the battery can be found by

$$i = V/R = 12V/0.92\ \Omega = 13A$$

25.5 RC Circuits (Resistors and Capacitors)

RC charging and discharging

a) Differential Equations

$$\varepsilon = R\frac{dq}{dt} + \frac{q}{C} \quad \text{(Charging)}$$

$$0 = R\frac{dq}{dt} + \frac{q}{C} \quad \text{(Discharging)}$$

b) Charge in the Capacitor

$$q = (C\varepsilon)(1 - e^{-t/RC}) \quad \text{(Charging)}$$

$$q = (C\varepsilon)e^{-t/RC} \quad \text{(Discharging)}$$

c) Current in the Resistor

$$i = \left(\frac{\varepsilon}{R}\right) e^{-t/RC} \quad \text{(Charging)}$$

$$i = -\left(\frac{\varepsilon}{R}\right) e^{-t/RC} \quad \text{(Discharging)}$$

where $e = 2.71828$ (Exponential Constant).

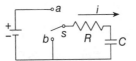

Figure 25.3 An RC Circuit

Problem Solving Example:

A resistor $R = 10$ megohms is connected in series with a capacitor $1 \ \mu F$. What is the time constant and half-life of this circuit?

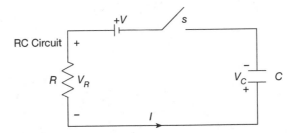

Suppose that we introduce an EMF in the circuit above, for example, by connecting a battery to the circuit. (See the figure.) In this case a current i will flow in the loop until C is charged up to the battery voltage V. Therefore, by Kirchoff's Voltage Law, V_R and i approach zero as V_C increases. The same current goes through R and C in the loop as we close the switch s;

$$i = \frac{V_R}{R} = \frac{dq}{dt} = C\frac{dV_C}{dt}$$

where q is the charge on the capacitor.

As a result of Kirchoff's second rule, $V_C + V_R = V$ and we obtain

$$RC\frac{dV_C}{dt} + (V_C - V) = 0$$

or

$$\frac{dV_C}{dt} + \frac{1}{RC}(V_C - V) = 0$$

The general solution of this differential equation is

$$V_C(t) = -Ae^{-t/\tau} + V \qquad (1)$$

where $\tau = RC$ is the time constant of the circuit and A is the integration constant. At $t = 0$, $V_C = 0$ because the voltage is initially only across R. Thus, substituting (1), $A = V$, and we get

$$V_C(t) = V\left(1 - e^{-t/\tau}\right)$$

Note that as t becomes very large

$$V_C(t \Rightarrow \infty) = V$$

The current i is

$$i(t) = C\frac{dV_C}{dt} = C\frac{d}{dt}\left[V\left(1 - e^{-t/\tau}\right)\right]$$

$$i(t) = \frac{CV}{\tau}e^{-t/\tau} = \frac{V}{R}e^{-t/\tau}$$

For the circuit under consideration,

$$\tau = RC = (10 \times 10^6\,\Omega)(10^{-6}\,\text{F}) = 10\,\text{s}$$

The half-life is the time required for the decay factor ($e^{-t/\tau}$) to be $1/2$; or

$$e^{t_h/\tau} = \frac{1}{2}$$

$$\ln e^{-t_h/\tau} = -t_h/\tau = \ell n \frac{1}{2}$$

$$t_h = -\tau \, \ell n \frac{1}{2}$$

$$= -(10 \text{ s})(\ell n \frac{1}{2}) = 6.9 \text{ s}$$

On the other hand, if $R = 10$ ohms, the time constant is only 10×10^{-6} s or 10 μs.

Quiz: Current and Resistance—Circuits

QUESTIONS 1–3 are related to the following diagram.

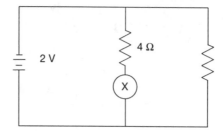

1. What is the total equivalent resistance of the circuit?

    (A) $\dfrac{1}{4}$ (D) 4

    (B) $\dfrac{1}{2}$ (E) 8

    (C) 2

2. How many total volts are dropped in the circuit?

    (A) $\dfrac{1}{4}$ (D) 4

    (B) $\dfrac{1}{2}$ (E) 8

    (C) 2

3. How many amps flow through the amp meter (X)?

(A) $\dfrac{1}{4}$

(D) 4

(B) $\dfrac{1}{2}$

(E) 8

(C) 2

4. What amount of current passes through the 15Ω resistor?

(A) 0.125 amp

(B) 0.417 amp

(C) 24 amps

(D) 8.0 amps

(E) 10.0 amps

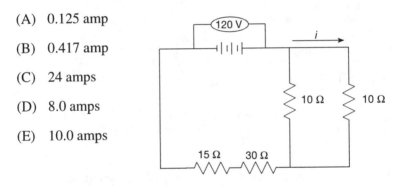

5. A load resistance of 5 kΩ can carry a maximum current of 10 amps. What is the power that is dissipated across the load resistance?

(A) 50 kW

(D) 500,000 W

(B) 500 W

(E) 500,000 kW

(C) 50,000 W

6. A 2.5 m long piece of wire with a radius of 0.65 mm has a resistance of 2Ω. If the length and the radius of the wire are doubled, what will be the new resistance?

(A) 16Ω (D) 2Ω

(B) 8Ω (E) 1Ω

(C) 4Ω

7. If ammeter A_3 reads 6 amps, ammeter A_1 will read how many amps?

(A) 2 amps (D) 5 amps

(B) 3 amps (E) 6 amps

(C) 4 amps

8. According to the laws of resistance, doubling the thickness of a given wire and making it 10 times longer will cause its resistance to be

(A) 40 times greater. (D) 2.5 times greater.

(B) 20 times greater. (E) 0.05 as much as before.

(C) 5 times greater.

9. As the diameter of a copper wire is decreased, its resistance

(A) remains constant. (D) varies with time.

(B) increases. (E) increases only with voltage.

(C) decreases.

10. A student has a battery and an assortment of resistors. How should he arrange the resistors in a circuit to obtain the maximum current flow when using all of the resistors?

(A) In a series

(D) All of these

(B) In parallel

(E) None of these

(C) In a series and parallel

ANSWER KEY

1.	(C)	6.	(E)
2.	(C)	7.	(C)
3.	(B)	8.	(D)
4.	(C)	9.	(B)
5.	(D)	10.	(B)

CHAPTER 26

The Magnetic Field

26.1 Force in a Magnetic Field

$$\overline{F}_B = q\overline{v} \times \overline{B}$$

$_B$ = force due to magnetic field

q = charge within the field

$\overline{v}$ = velocity of charge

$\overline{B}$ = magnetic field

Problem Solving Example:

Q An electron is projected into a magnetic field of flux density $B = 10$ Wb/m² with a velocity of 3×10^7 m/sec in a direction at right angles to the field. Compute the magnetic force on the electron and compare with the weight of the electron.

A The force on the electron in a magnetic field is given by $\overline{F} = q\,\overline{v} \times \overline{B}$, where is the velocity of the electron and $\overline{B}$ is the flux density of the magnetic field. In this case the velocity is perpendicular to the magnetic field so the force may be computed by a straightforward multiplication instead of taking the cross product.

The magnetic force is

$$F = qvB = (1.6 \times 10^{-19}\,\text{C})(3 \times 10^{7}\,\text{M/s})(10\,\text{Wb/m}^2) = 4.8 \times 10^{-11}\,\text{N}.$$

Where a weber per meter squared, Wb/m², is equivalent to a Newton second per coulomb meter.

$$F = mg = (9 \times 10^{-31}\,\text{kg})(9.8\,\text{m/s}^2) = 8.8 \times 10^{-30}\,\text{N}$$

The gravitational force is therefore negligible in comparison with the magnetic force.

26.2 Force on a Current-carrying Wire

$$\overline{F} = i\overline{\ell} \times \overline{B}$$

$\overline{F}$ = force due to magnetic field

i = current

$\overline{\ell}$ = length of wire and direction of wire

$\overline{B}$ = magnetic field

Problem Solving Example:

The figure shows a current of 25 amps in a wire 30 cm long and at an angle of 60° to a magnetic field of flux density 8.0×10^{-4} weber/m². What are the magnitude and direction of the force on this wire?

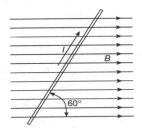

We must find the magnetic force on a wire placed in a magnetic field. The force on a differential element of the current carrying wire, is

$$d\overline{F} = i\,d\overline{\ell} \times \overline{B} \tag{1}$$

where i is the current in the wire, $d\overline{\ell}$ is a vector tangent to the wire in the direction of i, and $\overline{B}$ is the magnetic field. The net force on the wire is found by integrating $d\overline{F}$ over the length of the wire (this amounts to adding the contributions of each current element $d\overline{\ell}$ to the net force). Hence,

$$\overline{F} = \int_0^\ell i\,d\overline{\ell} \times \overline{B} \tag{2}$$

where ℓ is the length of the wire. Because i is constant and $\overline{B}$ is uniform (independent of position), we may rewrite (2) as

$$\overline{F} = i\left[\int_0^\ell d\overline{\ell}\right] \times \overline{B} \text{ or } \overline{F} = i\overline{\ell} \times \overline{B} \tag{3}$$

where $\overline{\ell}$ is a vector whose magnitude is the length of the wire, and whose direction is the direction of current flow in the wire. Now, in general, the magnitude of a cross product (such as $\left|\overline{\ell} \times \overline{B}\right|$) is defined as:

$$\left|\overline{\ell} \times \overline{B}\right| = \ell B \sin\theta$$

where ℓ is the magnitude of $\overline{\ell}$, B is the magnitude of $\overline{B}$, and θ is the angle between the directions of $\overline{\ell}$ and $\overline{B}$. Because we are asked to find the magnitude of the force, we have:

$$F = i\ell B \sin\theta \tag{4}$$

Substituting the given values into (4), we obtain:

$$F = (25\,\text{A})(.3\,\text{m})\left(8\times10^{-4}\,\frac{\text{Wb}}{\text{m}^2}\right)(\sin\,60°)$$

or
$$F = 5.2 \times 10^{-3} \text{N}$$

The wire is pushed away from the reader, from the stronger toward the weaker field.

26.3 Torque on a Current Loop

$$\bar{\tau} = NiA_T B \ \sin \theta$$

In vector notation

$$\bar{\tau} = \bar{\mu} \times \bar{B}$$

$\bar{\tau}$ = torque

N = number of loops (turns) in the coil

i = current

A_T = total area of the coil

$\bar{B}$ = magnetic field

θ = angle formed by a normal to the plane of the loop and the direction of $\bar{B}$

$\bar{\mu}$ = magnetic dipole moment = NiA_T

Problem Solving Example:

 A rectangular coil 30 cm long and 10 cm wide is mounted in a uniform field of flux density 8.0×10^{-4} N/Am. There is a current of 20 amps in the coil, which has 15 turns. When the plane of the coil makes an angle of 40° with the direction of the field, what is the torque tending to rotate the coil? (See figure on the following page.)

A The torque on a circuit in a field of magnetic induction, $\bar{B}$, is

$$\bar{\tau} = \bar{\mu} \times \bar{B} \tag{1}$$

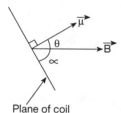

Plane of coil

where $\overline{\mu}$ is the magnetic moment of the circuit. (This is the property of the circuit that causes the torque to be exerted.) The magnitude of the magnetic moment is

$$\mu = NiA \qquad (2)$$

where N is the number of turns in the circuit, i is the current in the circuit, and A is the area it encloses.

The direction of $\overline{\mu}$ is given by the right-hand rule: wrap the fingers of your right hand around the circuit in the direction of the current, and the direction in which your thumb points will then be the sense of $\overline{\mu}$. Since we only want the magnitude of τ, we write

$$\tau = \mu B \ \sin \theta \qquad (3)$$

where τ, μ, B are the magnitudes of $\overline{\tau}$, $\overline{\mu}$, $\overline{B}$, and θ is the angle between the directions of $\overline{\mu}$ and $\overline{B}$ (see figure). Substituting (2) into (3), we obtain

$$\tau = NiAB \ \sin \theta \qquad (4)$$

However, the data is given in terms of flux density, not in terms of B. But flux density is actually equal to B because

$$\text{Flux density} = \frac{\phi}{A} = \frac{BA}{A} = B$$

where A is the area enclosed by the circuit, and ϕ is the flux cutting through the circuit. We still cannot proceed yet, because we do not have θ. The question gives us the angle between the plane of the coil and the direction of $\overline{B}$. (In the figure this is α.) The angle we need, θ, is $90° - \alpha = 50°$. Inserting the given data in (4), we find

$$\tau = \left[8 \times 10^{-4} \frac{N}{Am} \right] (15)(20\,A)(0.3\,m)(0.1\,m)(\sin 50°)$$

$$\tau = 0.0055 \text{ Nm}$$

26.4 Energy of a Dipole

$$U = -\overline{\mu} \cdot \overline{B}$$

U = magnetic energy

$\overline{\mu}$ = magnetic dipole moment

$\overline{B}$ = magnetic field

26.5 Circulating Charged Orbitals in a Magnetic Field

a) Radius of Path

$$r = \frac{mv}{qB}$$

r = radius of path

m = mass of charged particle

v = tangential velocity

q = charge of particle

B = magnetic field

b) Angular Velocity

$$\omega = \frac{v}{r} = \frac{qB}{m}$$

ω = angular velocity

v = tangential velocity

r = radius of path

q = charge of particle

B = magnetic field

m = mass of particle

c) Cyclotron Frequency of a Circulating Particle

$$\eta = \frac{qB}{2\pi m}$$

η = frequency

q = charge of particle

B = magnetic field

m = mass of particle

26.6 E/M for an Electron (Thomson's Experiment)

$$\frac{e}{m} = \frac{2yE}{B^2 \ell^2}$$

e = charge of electron

m = mass of electron

y = deflection of electron in a purely electric field

E = electric field

B = magnetic field

ℓ = length of deflecting plates

CHAPTER 27

Magnetic Fields and Currents

27.1 The Biot-Savart Law

$$d\overline{B} = \frac{\mu_0 i}{4\pi} \frac{\overline{d\ell} \times \overline{r}}{r^3}$$

B = magnetic field

μ_0 = permeability constant

i = current through a wire

ℓ = length of wire

r = distance from assumed point charge
     to a point in the magnetic field

θ = angle between r and the direction of the element

Note: $\mu_0 = 4\pi \times 10^{-7} \dfrac{T \bullet m}{A}$

Problem Solving Example:

A current of 30 amps is maintained in a thin, tightly wound coil of 15 turns, with a radius of 20 cm. What is the magnetic induction at the center of the coil?

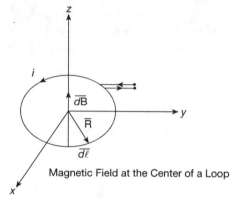

Magnetic Field at the Center of a Loop

The magnetic induction at the center of a coil of radius R with one turn can be found by using the Biot-Savart Law.

$$d\overline{B} = \frac{\mu_0 i}{4\pi} \frac{d\overline{\ell} \times \overline{R}}{R^3}$$

where the permeability constant

$$\mu_0 = 4\pi \times 10^{-7} \frac{\text{Tm}}{\text{A}}$$

and $d\overline{\ell}$ is an element of the coil in the direction of the current i. From the figure we have

$$d\overline{B} = \frac{\mu_0 i}{4\pi} \frac{d\overline{\ell} \overline{R} \sin 90°}{R^3} = \frac{\mu_0 i}{4\pi} \frac{d\ell}{R^2}$$

where $\overline{R}$ and $d\overline{\ell}$ are 90° apart. We have used the magnitude of $d\overline{B}$ because all the infinitesimal contributions to the magnetic induction from the infinitesimal lengths $d\overline{\ell}$ are in a direction perpendicular to the plane of the coil. Therefore, the total magnetic field at the center is the sum of the infinitesimal contributions dB or

$$B = \int dB = \frac{\mu_0 i}{4\pi} \frac{1}{R^2} \int d\ell = \frac{\mu_0 i}{4\pi} \frac{2\pi R}{R^2} = \frac{\mu_0}{2} \frac{i}{R}$$

The magnetic induction at the center of a coil containing N coils will be equal to the sum of the contributions due to each of the coils, or

$$B_T = NB = \frac{\mu_0 Ni}{2R} = (4\pi \times 10^{-7} \frac{Wb}{Am})(15)(30A)/2(0.2m)$$

$$= \left(4\pi \times 10^{-7} Wb/A\text{-}m\right) \frac{15 \times 30\,A}{2 \times 0.20\,m}$$

$$= 1.4 \times 10^{-3} Wb/m^2$$

The direction of B_T is perpendicular to the plane of the orbit.

27.2 Parallel Conductors

a) Force Between Two Parallel Conductors

$$\overline{F}_b = \frac{\mu_0 \ell i_b i_a}{2\pi d}$$

F_b = force by conductor a, on conductor b

μ_0 = permeability constant

ℓ = length of wire section

i_b = current through conductor b

i_a = current through conductor a

d = distance between conductors

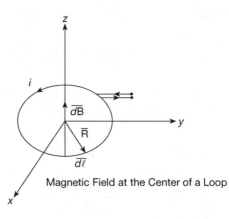

Magnetic Field at the Center of a Loop

Figure 27.1 Two Parallel Wires Carrying Parallel Currents

b)   Magnetic Field at a Distance r From a Wire Carrying Current

$$\overline{B}(r) = \frac{\mu_0 i}{2\pi r}$$

$\overline{B}(r)$ = magnetic field at a distance r from wire

μ_0 = permeability constant

r = distance from wire to point in field

Problem Solving Example:

Q   Two long, straight wires, each carrying a current of 9 A ˜^ the same direction, are placed parallel to each other. Find the force that each wire exerts on the other when the separation distance is 1×10^{-1} m.

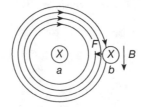

A The currents carried by the wires $i_a = 9$ A, $i_b = 9$ A, and the distance between them, $d = 1 \times 10^{-1}$ m, are given. The first current carrying wire a produces a field of induction B at all nearby points around it. The magnitude of B is

$$B = \frac{\mu_0}{2\pi} \frac{i_a}{d} \tag{1}$$

where i_a is the current through wire a, d is the distance separating the two wires, and the permeability constant $\mu_0 = 4\pi \times 10^{-7}$ W/(A · m). The other wire will find itself immersed in the field due to the first wire. The magnetic field it creates, that is its own self field, has no influence on its behavior. The magnetic force on this second current carrying wire (of length ℓ) is

$$\overline{F} = i_b \overline{\ell} \times \overline{B}$$

where i_b is the current through the second wire. Since the wire length is perpendicular to the magnetic field vector B, we then have

$$\overline{F} = i_b \ell B \tag{2}$$

F is directed inward toward the first wire. It is perpendicular to both B and to the length vector. Therefore, combining equations (1) and (2), we have

$$\frac{F}{\ell} = \frac{\mu_0}{2\pi} \frac{i_a i_b}{d}$$

The force per unit length on a wire is then

$$\frac{F}{\ell} = \frac{\mu_0}{2\pi} \frac{i_a i_b}{d} = \frac{4\pi \times 10^{-7} \frac{\text{Wb}}{\text{Am}}}{2\pi} \frac{i_a i_b}{d} = (2 \times 10^{-7} \frac{\text{Wb}}{\text{Am}}) \frac{(9\text{A})(9\text{A})}{0.1\text{m}}$$

$$= 1.6 \times 10^{-4} \frac{\text{N}}{\text{m}}$$

27.3 Ampere's Law

$$\oint \overline{B} \cdot \overline{d\ell} = \mu_0 i$$

$\overline{B}$ = magnetic field of wire

ℓ = length of wire

μ_0 = permeability constant

i = current

Problem Solving Example:

 A long, straight wire in a house carries an alternating current that has a maximum value of 20 amps. Calculate the maximum magnetic induction a distance 1 m away from the wire.

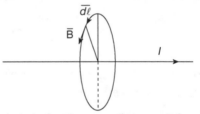

 The magnetic induction at a distance R from a wire containing a current i can be found by using Ampere's Law. Given a closed path c, and a current i through the closed curve, the following relation holds:

$$\int_c \overline{B} \cdot d\overline{\ell} = \mu_0 i$$

where the circle on the integral sign indicates that c is closed, and the permeability constant $\mu_0 = 4\pi \times 10^{-7}$ weber/amp-meter. Let the closed curve c of the above integral be in the shape of a circle of radius R, concentric with the current carrying wire as shown in the figure. Ampere's Law then becomes:

$$\int_c B \ \cos \theta \ d\ell = \int_c B \ d\ell$$

since $\overline{B}$ and $d\overline{\ell}$ are collinear and $\theta = 0°$. $\overline{B}$ is constant along the path c, therefore,

$$B\int d\ell = \mu_0 i$$
$$B \bullet 2\pi R = \mu_0 i$$

or

$$B = \frac{\mu_0 i}{2\pi R}$$

from which we have

$$B_{max} = \frac{\mu_0}{2\pi} \frac{i_{max}}{R} = \left(4\pi \times 10^{-7} \frac{Wb}{Am} \right) \left(\frac{1}{2\pi} \right) \left(\frac{20A}{1m} \right)$$

$$= 4 \times 10^{-6} \frac{Wb}{m^2}$$

This can be compared with the earth's magnetic field, which has an intensity of the order of 5×10^{-5} Wb/m^2.

27.4 Magnetic Field of a Solenoid

$$B_s = \mu_0 i_s N$$

B_s = magnetic field of solenoid

μ_0 = permeability constant

i_s = current of solenoid

N = number of turns in the coil

27.5 Magnetic Field of a Toroid

$$\overline{B}_T = \frac{\mu_0}{2\pi} \frac{i_T N}{r}$$

$\overline{B}_T$ = magnetic field of toroid

μ_0 = permeability constant

i_T = current of toroid

N = number of turns in the coil

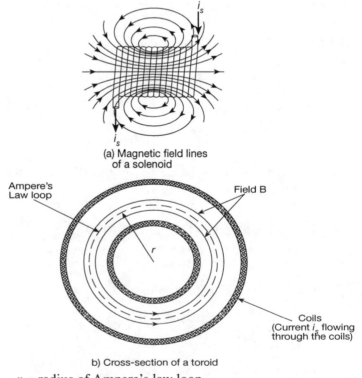

(a) Magnetic field lines
of a solenoid

Ampere's
Law loop

Field B

Coils
(Current i_s flowing
through the coils)

b) Cross-section of a toroid

r = radius of Ampere's law loop

Figure 27.2 Magnetic Fields of a Solenoid and a Toroid

27.6 Magnetic Field of a Current Loop

$$B(x) = \frac{\mu_0 i R^2}{2} \frac{I}{(R^2 + x^2)^{3/2}}$$

$B(x)$ = magnetic field of a current loop at a distance x along its axis

μ_0 = permeability constant

R = radius of current loop

x = distance along loop's axis

Problem Solving Example:

Use Ampere's Law to derive for the magnetic field of a toroid (N turns each carrying current i) of inner radius a and outer radius b at a distance r midway between a and b.

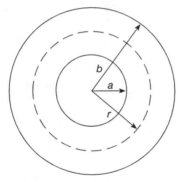

 We use Ampere's Law

$$\oint B \bullet d\ell = \mu_0 i_{in}$$

and take the Amperean path as a circle of radius r. Hence,

$$B(2\pi r) = \mu_0 Ni$$

$$B = \frac{\mu_0}{2\pi} \frac{Ni}{r} \quad r = \frac{a+b}{2}$$

$$B = \frac{\mu_0 Ni}{\pi(a+b)}$$

Quiz: The Magnetic Field—Magnetic Fields and Currents

QUESTIONS 1 and 2 refer to the following diagram and list of possible changes.

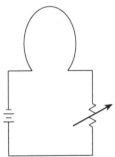

I. Decreases

II. Increases

III. Remains the same

IV. Equals zero

V. None of the above

1. As the voltage across the coil is increased, the magnetic field around the coil

    (A) I.   (D) IV.

    (B) II.   (E) V.

    (C) III.

2. If the coil is rotated in an external magnetic field and the voltage across the coil is decreased, the external force required to rotate the coil at a constant velocity

    (A) I.   (D) IV.

    (B) II.   (E) V.

    (C) III.

3. The energy required to operate a 100 watt lamp 5 hours a day for 30 days is

(A) 1.5 kw hr.

(D) 1,500 kw hr.

(B) 15 kw hr.

(E) 150 megawatt hr.

(C) 150 kw hr.

QUESTION 4 refers to the following diagram.

4. A wire in the plane of the page carries a current *I* directed toward the top of the page, as shown above. If the wire is located in a uniform magnetic field *B* directed out of the page, the force on the wire resulting from the magnetic field is

(A) directed into the page.

(B) directed out of the page.

(C) directed to the right.

(D) directed to the left.

(E) zero.

5. A magnet is dropped through an aluminum ring with the north pole of the magnet entering the ring first. What will be true of the induced magnetic field of the ring as the magnet passes through the ring?

    (A) There is no induced current in the ring so there is no induced magnetic field.

    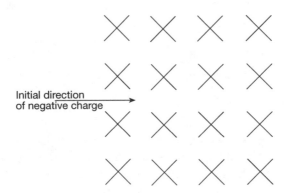

    (B) It will change from north up to north down.

    (C) It will change from south up to south down.

    (D) It will be constant with north up.

    (E) It will be constant with south up.

6. According to the following diagram, if a magnetic field is directed into this paper and a negatively charged particle is moving from left to right across it, there will be a force on the particle that pushes it

    X X X X

    X X X X

    Initial direction
    of negative charge →

    X X X X

    X X X X

(A)   into the page.

(B)   out from the page.

(C)   toward the bottom of the page.

(D)   toward the top of the page.

(E)   from left to right across the page.

7.   Which of the following diagrams show the direction of a magnetic field about a conducting wire?

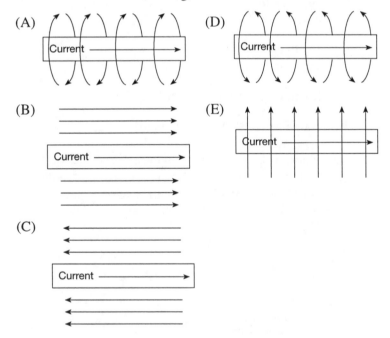

8.   In order for the torque on a current-carrying loop in a magnetic field to be at a minimum, the angle between the plane of the loop and the magnetic field must be

(A) 0°.   (B) 30°.

(C) 45°.   (D) 60°.

(E) 90°.

9.   A negative charge Q is fired into a uniform magnetic field B shown pointing into the page (by the ×'s). Which answer choices best describes its motion in the field?

B

(−)

Q

× × × × × × × × × × ×
× × × × × × × × × × ×
× × × × × × × × × × ×
× × × × × × × × × × ×
× × × × × × × × × × ×

(A) Straight   (D) Vibrates

(B) Parabolic   (E) Stops

(C) Circular

10. Conventional representation of magnetic field lines shows arrows

(A) surrounding a magnetic object.

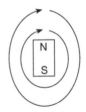

(B) emanating from a magnetic object.

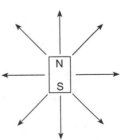

(C)   directed into a magnetic object.

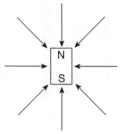

(D)   emanating from one pole of a magnetic object reaching around to another pole.

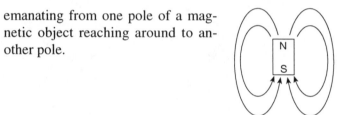

(E)   pointing from one pole on a magnetic object to the same pole on the nearest magnetic object.

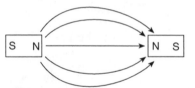

ANSWER KEY

1.	(B)	6.	(C)
2.	(E)	7.	(D)
3.	(B)	8.	(E)
4.	(C)	9.	(E)
5.	(B)	10.	(D)

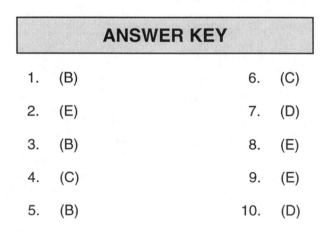

CHAPTER 28

Faraday's Law and Inductance

28.1 Magnetic Flux

$$\Phi_M = \int \overline{B} \cdot d\overline{S}$$

Φ_M = magnetic flux

$\overline{B}$ = magnetic field

$\overline{S}$ = surface area

28.2 Faraday's Law of Induction

$$\varepsilon = -N \frac{d\Phi_M}{dt}$$

ε = electromotive force

N = number of turns in the loop

Φ_M = magnetic flux

t = time

Problem Solving Example:

Q A coil of 600 turns is threaded by a flux of 8.0×10^{-5} weber. If the flux is reduced to 3.0×10^{-5} weber in 0.015 sec, what is the average induced EMF?

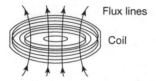

Flux lines

Coil

A We note that there is a change in magnetic flux. This immediately implies the use of Faraday's Law, which relates a change in magnetic flux to an induced ε (electromotive force). The flux linking the coil will induce an EMF in the coil. Faraday's Law states:

$$\varepsilon = \frac{-Nd\Phi_M}{dt} \qquad (1)$$

where N is the number of turns in the coil, and Φ_M is the magnetic flux linking the coil. We may write this using average values as

$$\varepsilon = -N\frac{\Delta\Phi_M}{\Delta t} \qquad (2)$$

where $\Delta\Phi_M$ is the change in magnetic flux over the interval Δt, and $\overline{EMF}$ is the average value of the resulting EMF. Substituting the values given in the statement of the problem into equation (2), we obtain

$$\varepsilon = -\frac{(600)\left(8 \times 10^{-5}\,\text{Wb} - 3 \times 10^{-5}\,\text{Wb}\right)}{(0.015\,\text{s})}$$

$$\varepsilon = -2\frac{\text{Wb}}{\text{s}}$$

But

$$1 \, \text{Wb} = 1 \, \frac{\text{Js}}{\text{C}}$$

Therefore,

$$\varepsilon = -2 \, \frac{\text{Js}}{\text{cs}} = -2\text{V}$$

28.3 Faraday's Law with Electric Fields

$$\oint \overline{\text{E}} \bullet \overline{d\ell} = \frac{-d\Phi_M}{dt}$$

$\overline{\text{E}}$ = electric field

$\overline{\ell}$ = length of wire

Φ_M = magnetic flux

t = time

28.4 Self-inductance

$$L = -\frac{\varepsilon}{di/dt}$$

L = self inductance

ε = electromotive force

i = current

t = time

Problem Solving Example:

Q A coil has a self-inductance of 1.26 millihenrys. If the current in the coil increases uniformly from zero to 1 amp in 0.1 sec, find the magnitude and direction of the self-induced EMF.

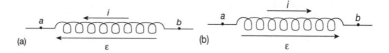

Direction of Self-induced EMF. (a) *i* increasing, ε opposite to *i*, point *a* at a higher potential than *b*. (b) *i* decreasing, ε and *i* in same direction, *b* at higher potential than *a*. (R = 0)

 The self-induced EMF on the inductance is given by

$$\varepsilon = -L\frac{di}{dt}$$

$$\varepsilon = \left(-1.26 \times 10^{-3}\,\text{henry}\right)\left(\frac{1\,\text{amp}}{.1\,\text{sec}}\right) \; .$$

Since 1 henry $= \dfrac{1 \;\; \text{volt} \bullet \sec}{\text{amp}}$

$$\varepsilon = \left(-1.26 \times 10^{-3}\,\frac{\text{volt} \bullet \sec}{\text{amp}}\right)\left(\frac{10\;\;\text{amp}}{\sec}\right)$$

$$= -12.6 \;\; \text{millivolts}$$

Since the current is increasing, the direction of this EMF is opposite to that of the current.

28.5 Inductance of a Solenoid

$$L = \frac{N\Phi_M}{i} = \mu_0 n^2 \ell A$$

L = inductance

N = number of turns in the coil

Φ_M = magnetic flux

i = current

n = number of turns per unit length

ℓ = length of the solenoid

A = cross-sectional area of the solenoid

μ_0 = permeability of the core

28.6 The RL Circuit

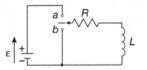

Figure 28.1 RL Circuit

28.7 Current Change in an RL Circuit

$$i = \frac{\varepsilon}{R}\left(1 - e^{\frac{-tR}{L}}\right)$$

i = current

ε = electromotive force

R = resistance

L = inductance

t = time

with
$$\frac{R}{L} = \frac{1}{\tau} = \frac{1}{\text{time constant}}$$

Note: Decay of current in an RL circuit may be expressed as

$$i = \frac{\varepsilon}{R} e^{\frac{-tR}{L}}$$

Problem Solving Example:

 An inductor of inductance 3 henrys and resistance 6 ohms is connected to the terminals of a battery of EMF 12 volts and of negligible internal resistance. a) Find the initial rate of increase of current in the circuit. b) Find the rate of increase of current at the instant when the current is 1 ampere. c) What is the instantaneous current 0.2 sec after the circuit is closed?

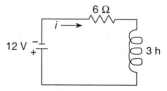

 a) Using Kirchoff's voltage law around the circuit yields

$$V = Ri(t) + L\frac{di(t)}{dt}$$

or
$$\frac{di(t)}{dt} = \frac{V}{L} - \frac{R}{L}i(t) \tag{1}$$

Here, $L\dfrac{di}{dt}$ is the EMF induced in the inductance due to the changing current. Equation (1) relates the rate of change in current at a time t to the current at time t. At $t = 0$, the initial current is zero. Hence, the initial rate of increase of current is

$$\frac{di}{dt} = \frac{V}{L} = \frac{12V}{3 \, h} = 4 \, \text{amp/sec}$$

b) When $i = 1$ amp,

$$\frac{di}{dt} = \frac{12V}{3 \, h} - \frac{6\Omega}{3 \, h} \times 1 = 2 \, \text{amp/sec}$$

c) The current at any time can be found by solving equation (1) for i. The function

$$i = \frac{V}{R}\left(1 - e^{-\frac{R}{L}t}\right)$$

can be shown by substitution into the differential equation to satisfy both equation (1) and the initial condition $i = 0$ at $t = 0$. At $t = 0.2$ sec, we then have

$$i = \frac{V}{R}\left(1 - e^{-\frac{Rt}{L}}\right) = \frac{12V}{6\Omega}\left(1 - \varepsilon^{-(6\Omega)(0.25)/3H}\right) = 2\left(1 - \varepsilon^{-0.4}\right)\text{amps}$$

$$= 2(1 - .670)\,\text{A} = 0.659 \, \text{A}$$

28.8 Energy Stored by an Inductor

$$U = \frac{1}{2}Li^2$$

U = stored energy

L = inductance

i = current

28.9 Energy Stored in a Magnetic Field

$$u = \frac{B^2}{2\mu_0}$$

u = stored energy

B = magnetic field

μ_0 = permeability constant

28.10 EMF Due to Mutual Induction

$$\varepsilon = -M\frac{di}{dt} \qquad \text{(For each coil)}$$

ε = electromotive force

M = coefficient of mutual inductance

28.11 Mutual Inductance of Close-packed Coils (i.e., Solenoids, Toroids, etc.)

$$M = \frac{N\Phi_M}{i}$$

M = coefficient of mutual induction

N = number of turns in coil

Φ_M = magnetic flux

i = current

Problem Solving Example:

A long solenoid of length ℓ and cross-sectional area A is closely wound with N_1 turns of wire. A small coil of N_2 turns surrounds it at its center, as in the figure. Find the mutual inductance of the coils.

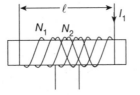

 If a current i in a coil is changing at a rate $\dfrac{di}{dt}$, then a voltage across the coil will be induced in the direction to oppose the change,

$$V_{\text{coil}} = -L\frac{di}{dt}$$

where L is the self-inductance of the coil. Similarly, if the time varying magnetic flux of one coil links another coil, then it will induce a voltage in the second coil equal to

$$V'_{\text{coil}} = -M\frac{di}{dt} \qquad (1)$$

where I is the current in the first coil. The constant M is the mutual inductance of two coils. Its value is determined only by the geometry and positioning of the coils. In our problem, the magnetic field in the solenoid (away from the edges) is given by

$$B = \frac{\mu_0 N_1 i_1}{\ell}$$

where I_1 is the current in the solenoid coil. The flux Φ through the N_2 coil is BA, where A is the area of the second coil. Then the voltage induced in the second coil by I_1 is, by Faraday's Law,

$$V_{12} = -N_2\frac{d\Phi}{dt} = -\frac{\mu_0 N_1 N_2 A}{\ell}\frac{di_1}{dt}$$

Therefore, we see by comparison with (1) that the mutual inductance of this system is

$$M = \frac{\mu_0 A N_1 N_2}{\ell}.$$

If $\ell = 0.50$ m, $A = 10$ cm$^2 = 10^{-3}$ m^2, $N_1 = 1{,}000$ turns, $N_2 = 10$ turns, then

$$M = \frac{\left(4\pi \times 10^{-7} \ \text{henry/m}\right)\left(10^{-3} \text{m}^2\right)(1{,}000 \ \text{turns})(10 \ \text{turns})}{(0.5\,\text{m})}$$

$$\approx 25 \times 10^{-6} \ \text{henry} \approx 25 \ \mu\text{H}$$

Quiz: Faraday's Law and Inductance

1. A battery is connected to a light bulb through a switch and transformer as shown below. Which one of the following statements best describes what happens when the switch is opened and closed?

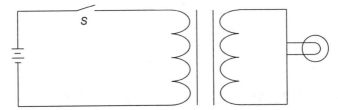

(A) As long as the switch is closed the bulb will light.

(B) The bulb will light momentarily when the switch is closing and then go out.

(C) The bulb will light momentarily when the switch is opened.

(D) The bulb will light momentarily when the switch is closing and momentarily again when it's opened.

(E) The bulb will never light while wired in this arrangement.

2. The magnetic flux through a wire loop

(A) requires a time-varying magnetic field.

(B) is maximum when the plane of the loop is perpendicular to the field.

(C) increases when the magnetic field is decreased.

(D) increases when the magnetic field is zero.

(E) None of these.

3. A coil of wire is connected to a 65-volt battery as shown. The wire produces 1,000 calories per second. Which one of the following is *closest* to the resistance of the coil?

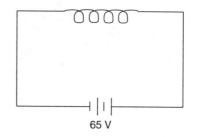

65 V

(A) 1 ohm   (D) 1,000 ohms

(B) 10 ohms   (E) None of the above.

(C) 100 ohms

4. An induced current in a wire loop

(A) results from a constant magnetic flux.

(B) can result only from a time-varying magnetic field.

(C) is in such a direction that its effects oppose the change producing it.

(D) always decreases when the magnetic flux decreases.

(E) None of the above.

5. A metal rod is pulled along a metal frame (which forms a rectangular loop with the rod) with a constant velocity. With a uniform magnetic field perpendicular to the cross-section of the frame, the voltage induced across the rod

(A) gives rise to a current that augments the field.

(B) is inversely proportional to the magnetic field.

(C) is called motional emf.

(D) is independent of the velocity of the rod.

(E) None of the above.

QUESTIONS 6 and 7 refer to the following diagram:

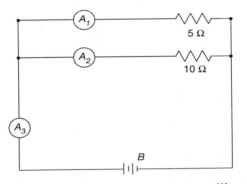

6.   If ammeter A_3 reads 6 amps, ammeter A_1 will read how many amps?

(A) 2 amps   (D) 5 amps

(B) 3 amps   (E) 6 amps

(C) 4 amps

7.   According to the laws of the resistance, doubling the thickness of a given wire and making it 10 times longer will cause its resistance to be

(A) 40 times greater.   (D) 2.5 times greater.

(B) 20 times greater.   (E) 0.05 as much as before.

(C) 5 times greater.

8. The primary coil of an induction coil has a battery and switch in series with it. The secondary coil has a galvanometer in series with it as shown. Consider the following situations:

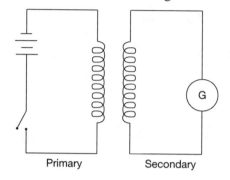

Primary   Secondary

    I. The switch is closing.

    II. The switch is held closed.

    III. The switch is opening.

    IV. The switch is held opened.

During which situation(s) will the galvanometer read?

(A) II only.   (D) I only.

(B) III only.   (E) I and III only.

(C) II and IV.

9. In a step-up transformer, if there are 20 turns on the secondary for each turn on the primary, then a primary voltage of 110 volts becomes

(A) 5.5 V.   (D) 2,200 V.

(B) 110 V.   (E) 11,000 V.

(C) 130 V.

10. If the rate of change of the current in an inductor is tripled, the induced EMF in the inductor is changed by a factor of

(A) $\dfrac{1}{9}$.

(D) 3.

(B) $\dfrac{1}{3}$.

(E) 9.

(C) 1.

ANSWER KEY

1.	(D)	6.	(C)
2.	(E)	7.	(D)
3.	(A)	8.	(E)
4.	(C)	9.	(D)
5.	(C)	10.	(D)

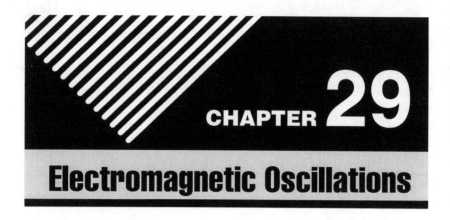

CHAPTER **29**

Electromagnetic Oscillations

29.1 Energy Transfers in LC Oscillations

a) Energy Stored in a Capacitor

$$U_E = \frac{1}{2}\frac{q^2}{C}$$

U_E = Energy Stored Due to the Electric Field of the Capacitor

q = Charge

C = Capacitance

b) Energy Stored in an Inductor

$$U_M = \frac{1}{2}Li^2$$

U_M = Energy Stored Due to the Magnetic Field of the Inductor

L = Inductance

i = Current

c) Total Energy

$$U_T = U_E + U_M = \frac{1}{2}\frac{q^2}{C} + \frac{1}{2}Li^2$$

29.2 Frequency of LC Oscillations

$$\omega = \frac{1}{\sqrt{LC}}$$

ω = Angular Frequency

L = Inductance

C = Capacitance

29.3 Conservation of Energy for LC Oscillations

Differential Equation

$$L\frac{d^2q}{dt^2} + \frac{q}{C} = 0$$

where q is the electric charge and t is time.

29.4 Conservation of Energy for Damped (RCL) Oscillations

$$L\frac{d^2q}{dt^2} + R\frac{dq}{dt} + \frac{q}{C} = 0$$

L = inductance

q = charge

t = time

R = resistance

C = capacitance

29.5 Solution for the Energy Conservation Differential Equations of RCL Oscillations

$$q = q_m e^{\frac{-Rt}{2L}} \cos(\omega t + \phi)$$

q = charge

q_m = charge amplitude

R = resistance

t = time

L = inductance

ω = angular frequency

ϕ = phase constant

29.6 Forced Oscillations

a) With Impressed EMF,

$$\varepsilon = \varepsilon_m \sin \omega_e t$$

The differential equation becomes,

$$L\frac{d^2q}{dt^2} + R\frac{dq}{dt} + \frac{q}{C} = \varepsilon_m \sin \omega_e t$$

where ε is the EMF, ε_m is the EMF amplitude, and ω_e is the angular frequency of the EMF.

b) Equation for Current

$$i = i_m \sin(\omega_e t + \phi)$$

where i is the current and i_m is the current amplitude.

Note: i_m has a maximum value when $\omega_e = \omega$; this condition is called resonance.

Problem Solving Examples:

 Consider the coupled inductor-capacitor circuit shown below. Determine the ratio of the frequency of the anti-symmetric mode to that of the symmetric mode ω_a/ω_s. Let $k = 1/LC$ and $\kappa = 1/L\gamma$.

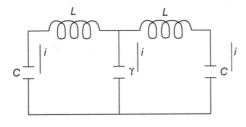

 By Kirchhoff's current law

$$i_1 = i + i_2 \Rightarrow i = i_1 - i_2$$

By Kirchhoff's voltage laws:

$$\frac{-q_1}{c - Li'_1} - \frac{q_1 - q_2}{\gamma} = 0$$

$$\frac{-q_2}{c - Li'_2} + \frac{q_1 - q_2}{\gamma} = 0$$

Differentiate and let $k = 1/LC$ and $\kappa = 1/L\gamma$. Then we obtain

$$i_1'' = -i_1 k + (i_2 - i_1)\kappa$$
$$i_2'' = -i_2 k - (i_2 - i_1)\kappa$$

Add and subtract the equations letting $y = i_1 + i_2$ and $z = i_2 - i_1$, respectively, to get

$$y'' = -ky \text{ and } z'' = -kz = 2\kappa z.$$

Hence, the ratio

$$\frac{\omega_a}{\omega_s} = \sqrt{\frac{(k+2\kappa)}{k}} = \sqrt{1 + \frac{2\kappa}{k}}$$

 An RLC circuit vibrates subject to the initial conditions $I = I_0$ and $I' = 0$ at $t = 0$. What is the time dependent current in the critical damping case? Let $\gamma = R/L$.

$$\omega_0^2 = \frac{1}{LC}, \quad \omega = \sqrt{\frac{\omega_0^2 - \gamma^2}{4}} \text{ and } \tan = \frac{-\gamma}{2\omega}$$

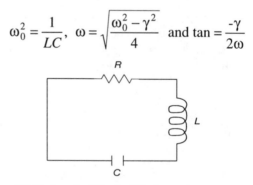

 For an RLC circuit, Kirchoff's law gives

$$-RI - LI - \frac{Q}{C} = 0$$

or differentiating and defining

$$\omega_0^2 = \frac{1}{LC} \text{ and } \gamma = \frac{R}{L}$$
$$I'' + \gamma I' + \omega_0^2 I = 0$$

for the critical dumping case,

$$\omega_0{}^2 = \frac{\gamma^2}{4}$$

and the solution is

$$I = (A + Bt)e^{-\gamma\, t/2}$$

If $I = I_0$ and $I' = 0$ at $t = 0$, then the desired solution is

$$I = I_0(1 + \frac{\gamma}{2}t)e^{-\gamma\, t/2}$$

Evaluate the circuit shown below to determine the anti-symmetric mode frequency. Let $k = 1/LC$, $\kappa_1 = 1/L\gamma$, and $\kappa_2 = \beta/L$.

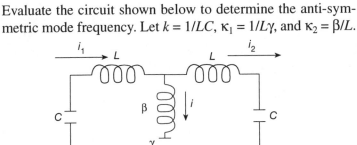

By Kirchoff's current law

$$i = i_1 + i_2$$

According to Kirchoff's voltage law:

$$-\frac{q_1}{C} Li'_1 - \beta i' - \frac{(q_1 - q_2)}{\gamma} = 0$$

$$-\frac{q_2}{C} - Li'_2 + \frac{(q_1 - q_2)}{\gamma} + \beta i' = 0$$

Differentiate and set

$$\kappa = \frac{1}{LC}, \quad \kappa_1 = \frac{1}{L\gamma}, \quad \kappa_2 = \frac{\beta}{L}$$

Note that κ_2 is dimensionless. Then we find

$$i_1'' (1+\kappa_2) - \kappa_2 i_2'' + (k+\kappa_1)i_1 - \kappa_1 i_2 = 0$$
$$i_2'' (1+\kappa_2) - \kappa_2 i_1'' + (k+\kappa_1)i_2 - \kappa_1 i_1 = 0$$

Add and subtract the equations letting

$$y = i_1 + i_2 \quad \text{and} \quad z = i_2 - i_1$$

to obtain

$$y'' + ky = 0 \quad \text{and} \quad z'' + \frac{k+2\kappa_1}{1+2\kappa_2} z = 0$$

thus

$$w_a = \sqrt{\frac{(k+2\kappa_1)}{(1+2\kappa_2)}}$$

CHAPTER 30

Alternating Currents

30.1 A Resistive Circuit

Figure 30.1 Single-loop Resistive Circuit with an AC Generator

a) Voltage

$$V_R = V_{R,m} \sin \omega t$$

b) Current

$$i_R = \frac{V_R}{R} = \left(\frac{V_{R,m}}{R} \right) \sin \omega t$$

where

$$i_R = i_{R,m} \sin \omega t$$

V_R = voltage

$V_{R,m}$ = maximum voltage

ω = angular frequency

t = time

i_R = current

$i_{R,m}$ = maximum current

Problem Solving Example:

 What is the impedance of a 1-henry inductor at angular frequencies of 100, 1,000, and 10,000 rad/sec?

 The circuit element is an inductor; therefore, the impedance is purely reactive.

At an angular frequency of 1,000 rad/sec, the reactance of a 1-henry inductor is

$$X_L = \omega L = 10^3 \frac{\text{rad}}{\text{sec}} \times 1\,\text{H} = 1,000\ \text{ohms}$$

At a frequency of 10,000 rad/sec the reactance of the same inductor is 10,000 ohms, while at a frequency of 100 rad/sec it is only 100 ohms.

30.2 A Capacitive Circuit

Figure 30.2 Single-loop Capacitive Circuit with an AC Generator

a) Voltage

$$V_c = V_{c,m} \sin \omega t$$

b) Capacitive Reactance

$$X_c = \frac{1}{\omega C}$$

c) Current

$$i_c = \left[\frac{V_{c,m}}{X_c} \right] \sin(\omega t + 90°) = i_{c,m} \sin(\omega t + 90°)$$

V_c = voltage

$V_{c,m}$ = maximum voltage

ω = angular frequency

t = time

X_c = capacitive reactance

i_c = current

$i_{c,m}$ = maximum current

C = capacitance

Problem Solving Example:

A capacitor is found to offer 25 ohms of capacitive reactance when connected in a 400-cps circuit. What is its capacitance?

Capacitive resistance, X_C, is given by

$$X_C = \frac{1}{2\pi f C}$$

Here X_C = 25 ohms and f = 400 cps. Then

$$X_C = 25 \text{ ohms} = \frac{1}{2\pi \times 400 \text{ cps} \times C}$$

or

$$C = \frac{1}{2\pi \times 400 \text{ cps} \times 25 \text{ ohms}}$$

Since
$$1 \text{ farad} = 1\frac{\text{coul}}{\text{volt}} = \frac{1 \text{ coul}}{\text{amp} \cdot \text{ohm}} = 1\frac{\text{sec}}{\text{ohm}}$$

$$C = \frac{1}{2\pi \times 400 \times 25} \text{ farad}$$

$$= 1.6 \times 10^{-5} \text{farad} = 16\mu\text{F}$$

30.3 An Inductive Circuit

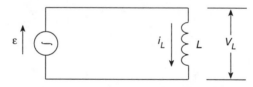

Figure 30.3 Single-loop Inductive Circuit with an AC Generator

a) Voltage

$$V_L = V_{L,m} \sin \omega t$$

b) Inductive Reactance

$$X_L = \omega L$$

c) Current

$$i_L = \left[\frac{V_{L,m}}{X_L}\right] \sin(\omega t - 90°) = i_{L,m} \sin(\omega t - 90°)$$

V_L = voltage

$V_{L,m}$ = maximum voltage

ω = angular frequency

t = time

X_L = inductive reactance

L = inductance

i_L = current

$i_{L,m}$ = maximum current

Problem Solving Example:

Q A coil having resistance and inductance is connected in series with an AC ammeter across a 100-volt DC line. The meter reads 1.1 amperes. The combination is then connected across a 110- volt AC 60 cycle line and the meter reads .55 ampere. What are the resistance, the impedance, the reactance, and the inductance of the coil?

A Note that with DC by Ohm's law,

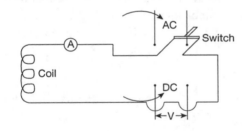

$$I = \frac{V}{R} \qquad R = \frac{V}{I}$$

$$R = \frac{100 \text{ V}}{1.1 \text{ A}} = 90.9 \, \Omega$$

(resistance)

On AC, however, the reactance of the inductance must also be taken into account (it is zero for DC current). Ohm's law still applies with the impedance taking the place of resistance. Therefore,

$$\frac{E}{I} = Z = \frac{110\,V}{.55\,A} = 200 \quad \text{ohms (impedance)}$$

But
$$\bullet Z = \sqrt{R^2 + X^2}$$

where X is the reactance of the inductance.

$$= 100^2 + X^2 = (200 \quad \Omega)^2$$

therefore

$$X^2 = (40,000 - 10,000)\Omega^2 = 30,000 \quad \Omega^2$$

$$X = \sqrt{30,000 \quad \Omega^2} = 173 \quad \Omega \text{ (reactance)}$$

And
$$X = 2\pi nL$$

where n is the number of times per second the current alternates. For DC current, $n = 0$, and the inductance has no effect on the circuit. (The effect of the inductance on the behavior of the circuit increases appreciably as the frequency of the AC current increases.)

Therefore,

$$L = \frac{X}{2\pi n} = \frac{173\Omega}{2\pi(60/s)} = .459 \quad \text{henry (inductance)}$$

30.4 Current Amplitude

$$i_m = \frac{\varepsilon_m}{\sqrt{R^2 + (X_c - X_L)^2}}$$

i_m = current amplitude

ε_m = electromotive force amplitude

R = resistance

X_c = capacitive reactance

X_L = inductive reactance

Problem Solving Example:

 When a 2-V cell is connected in series with two electrical elements, the current in the circuit is 200 mA. If a 50-cycle $\cdot$ s^{-1}, 2-V AC source replaces the cell, the current becomes 100 mA. What are the values of the circuit elements? Suppose that the frequency is increased to 1,000 cycles $\cdot$ s^{-1}. What is the new value of the current?

A Since a DC current can flow through the elements, neither of them can be a capacitor, which offers an infinite resistance to direct current. The resistance of the elements is

$$R = \frac{V}{I} = \frac{2 \text{ V}}{200 \times 10^{-3} \text{ A}} = 10 \text{ }\Omega.$$

Since the current changes when alternating current is supplied, the current must also contain an inductor of inductance L. The reactance (the *ac* analogue of resistance) of an inductor is $X = \omega L$ where $\omega = 2\pi n$. n is the frequency of variation of voltage. The impedance Z (i.e., the ratio of V to I at a set of terminals of a circuit as shown in Figure (a)) of the series combination of the resistor and inductor (Figure (b)) is

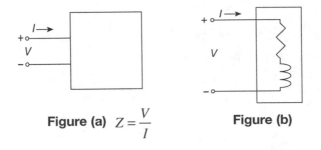

Figure (a) $Z = \dfrac{V}{I}$ **Figure (b)**

$$Z^2 = R^2 + X^2$$

$$Z = \sqrt{R^2 + \omega^2 L^2} = \frac{V}{I} = \frac{2 \text{ V}}{100 \times 10^{-3} \text{ A}} = 20 \ \Omega$$

$$\therefore \quad 100 \ \Omega^2 + \omega^2 L^2 = 400\Omega^2 \text{ or } L^2 = \frac{300 \ \Omega^2}{\omega^2}$$

$$\therefore \quad L = \frac{17.32 \ \Omega}{2\pi \times 50 \text{ s}^{-1}} = 0.055 \text{ H}$$

When the frequency is increased to 10^3 cycles $\cdot$ s^{-1},

$$I' = \frac{V'}{Z'} = \frac{2 \text{ V}}{\sqrt{100 \ \Omega^2 + (2\pi \times 10^3 \times 0.055)^2 \Omega^2}} = \frac{2}{346} \text{ A}$$

$$= 5.78 \text{ mA}$$

We see that when the frequency $n = 0$, $I = 200$ mA. When $n = 50$ sec^{-1}, $I = 100$ mA, and when $n = 1,000$ sec^{-1}, $I = 5.77$ mA. The inductor does then act as a resistor. Its opposition to current (the reactance) is directly proportional to frequency as is seen by the relation $X = (L)\omega = (2\pi L)n = Kn$ where K is a constant.

30.5 Impedance

$$Z = \sqrt{R^2 + (X_c - X_L)^2}$$

z = impedance

X_c = capacitive reactance

X_L = inductive reactance

Problem Solving Example:

What is the impedance of a 1-µF capacitor at angular frequencies of 100, 1,000, and 10,000 rad/sec?

The circuit element is a capacitor; therefore, the impedance is purely reactive.

At an angular frequency of 1,000 rad/sec, the reactance of a 1-µF capacitor is

$$X_c = \frac{1}{\omega C} = \frac{1}{\left(10^3 \, \text{rad/sec}\right) \times 10^{-6} \, \text{F}} = 1,000 \text{ ohms}$$

At a frequency of 10,000 rad/sec the reactance of the same capacitor is only 100 ohms, and at a frequency of 100 rad/sec it is 10,000 ohms.

30.6 Phase Angle

$$\tan \phi = \frac{X_c - X_L}{R}$$

ϕ = phase angle

X_c = capacitive reactance

X_L = inductive reactance

R = resistance

Problem Solving Example:

In the circuit of Figure (a), on next page, the values are as follows: $C = 30$ µF, $V = 120$ volts, and $R = 25$ ohms. The voltage is alternating with frequency $f = 60$ cycles/sec. What is the current? What is the phase angle?

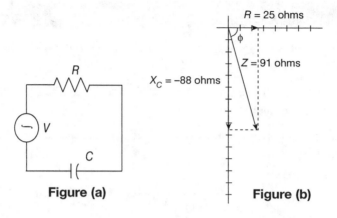

Figure (a)   Figure (b)

 The reactance of the capacitor depends on frequency.

$$X_c = \frac{+1}{2\pi f C} = \frac{+1}{(2\pi)(60/s)(30 \times 10^{-6} F)} = +88\,\Omega$$

The impedance Z is the vector addition of the reactance and pure resistance, since the reactance of the capacitor causes the current to lag the voltage by a phase of 90°.

$$Z = \sqrt{R^2 + X_c^2} = \sqrt{25^2 + 88^2} = 92\,\Omega$$

$$I = \frac{V}{Z} = \frac{120 \text{ volts}}{91 \text{ ohms}} = 1.3 \text{ amp}$$

The phase angle ϕ is the angle between the impedance and the pure resistance of the circuit (see Figure (b)).

$$\cos\phi = \frac{R}{Z}$$

$$\phi = \cos^{-1}\frac{R}{Z} = \cos^{-1}\frac{25 \text{ ohms}}{92\,\Omega} = +74.2°$$

30.7 Resonance

The maximum value of current, i_m, occurs when

$$\omega = \frac{1}{\sqrt{LC}}$$

ω = angular frequency

L = inductance

C = capacitance

30.8 Power: Two Equations

a)
$$P = \left(\frac{i_m}{\sqrt{2}}\right)^2 R$$

or

b)
$$P = \left(\frac{\varepsilon_m}{\sqrt{2}}\right)\left(\frac{i_m}{\sqrt{2}}\right)\cos\phi$$

P = average power

i_m = current amplitude

R = resistance

ε_m = electromotive force amplitude

ϕ = phase angle

30.9 The Transformer

a) Voltage Transformation

$$V_2 = V_1\left(\frac{N_2}{N_1}\right)$$

V = voltage

N = number of turns of wire

b) Current Transformation

$$i_2 = i_1\left(\frac{N_1}{N_2}\right)$$

i = current

N = number of turns of wire

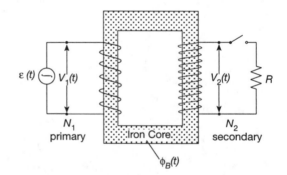

Figure 30.4 Ideal Transformer—Two Coils Wound on a Soft Iron Core

Problem Solving Example:

Q   In a step-up transformer, if there are 20 turns on the secondary for each turn on the primary, what does the primary voltage of 110 volts become?

A The ratio of the number of turns is the same ratio as the voltage. The primary of 110 volts is to one turn as the new voltage is to 20 turns. Thus, 20 times 110 produces 2,200 volts on the secondary.

Quiz: Electromagnetic Oscillations—Alternating Currents

1. The impedance of an RLC circuit

    (A) has the units of henrys.

    (B) is independent of the voltage frequency.

    (C) does not affect the power dissipated in the circuit.

    (D) is a minimum at the resonant frequency of the circuit.

    (E) is a maximum at the resonant frequency of the circuit.

2. In order for the torque on a current-carrying loop in a magnetic field to be at a minimum, the angle between the plane of the loop and the magnetic field must be

    (A) 0°.   (D) 60°.

    (B) 30°.   (E) 90°.

    (C) 45°.

3. If the switch in this circuit is closed, what will be the change in power?

    (A) 48 W decrease

    (B) 48 W increase

    (C) 72 W increase

    (D) 24 W decrease

    (E) 24 W increase

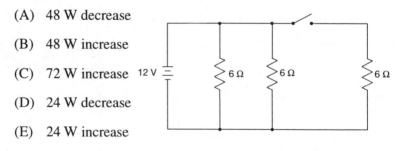

4. In the circuit shown, when switch 1 is closed, the instantaneous current in the circuit (assuming the resistance of the inductor to be negligible) is

   (A) zero.

   (B) V/R.

   (C) V/L.

   (D) $(V/L) \, dt$.

   (E) $(V/R) \, dt$.

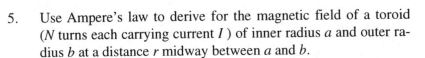

5. Use Ampere's law to derive for the magnetic field of a toroid (N turns each carrying current I) of inner radius a and outer radius b at a distance r midway between a and b.

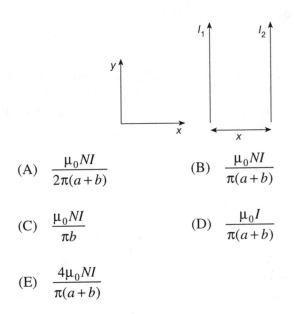

   (A) $\dfrac{\mu_0 NI}{2\pi(a+b)}$

   (B) $\dfrac{\mu_0 NI}{\pi(a+b)}$

   (C) $\dfrac{\mu_0 NI}{\pi b}$

   (D) $\dfrac{\mu_0 I}{\pi(a+b)}$

   (E) $\dfrac{4\mu_0 NI}{\pi(a+b)}$

6. In the circuit shown, the battery has an EMF and internal resistance r. The meter reads 12.0 volts when the switch is open. When the switch is closed, the steady-state reading on the voltmeter is 11.6 volts. The resistance of the wires and switch are negligible. $R_{meter} = \infty$. What is the internal resistance (r) of the battery?

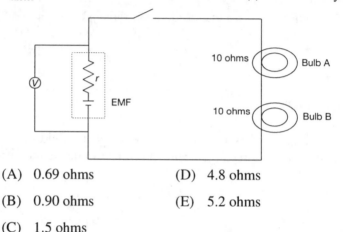

(A) 0.69 ohms   (D) 4.8 ohms

(B) 0.90 ohms   (E) 5.2 ohms

(C) 1.5 ohms

7. Consider that a sliding conductive bar closes the circuit shown below and moves to the right with a speed $v = 4$ m/s. If $l = 1.5$ m, $R = 12\Omega$, and $B = 5$ T, then find the magnitude of the induced power and the direction of the induced current.

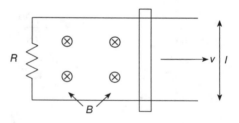

(A) 75 W, counterclockwise

(B) 75 W, clockwise

(C) 2.5 W, counterclockwise

(D) 2.5 W, clockwise

(E) 0 W, there is no current flow

8. In the circuit below, switch 1 is closed until the capacitor is fully charged. Then, switch 1 is opened and switch 2 is closed. Immediately after switch 2 is closed, the instantaneous current through the resistance R is

(A) 0.

(B) $\dfrac{V}{R}$.

(C) $\dfrac{R}{V}$.

(D) CV.

(E) CR.

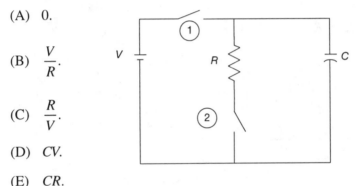

9. Consider a circuit that consists of four resistors (each with $R = 1$ $M\Omega$), a capacitor ($C = 1$ mF), and a battery ($V = 10$ MV) as shown. If the capacitor is fully charged and then the battery is removed, find the current at $t = 0.5$ s as the capacitor discharges.

(A) 40 A

(B) 20 A

(C) 24.3 A

(D) 14.7 A

(E) 5.4 A

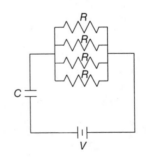

10. Recall the equation for a series RLC circuit. Compare this to the parallel resonant circuit shown and find R_p if a series RLC circuit and the parallel RLC circuit are to have the same equations for the potential of capacitor while they both have the same L, C, and Q.

(A) $R_p = R$

(B) $R_p = L$

(C) $R_p = \dfrac{L}{C}$

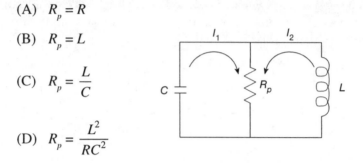

(D) $R_p = \dfrac{L^2}{RC^2}$

(E) $R_p = \dfrac{L}{RC}$

ANSWER KEY			
1.	(D)	6.	(A)
2.	(D)	7.	(A)
3.	(E)	8.	(B)
4.	(A)	9.	(E)
5.	(B)	10.	(E)

CHAPTER 31

Electromagnetic Waves

31.1 Speed of Light

$$c = \frac{1}{\sqrt{\varepsilon_0 \mu_0}} \quad \text{units: meters/sec}$$

c = speed of light (electromagnetic radiation) in free space

ε_0 = electric permittivity constant

μ_0 = magnetic permeability constant

31.2 The Poynting Vector

a) Definition

The rate of energy transport per unit area in an electromagnetic wave may be described as the Poynting vector $\overline{S}$.

b) Vector Equation

$$\overline{S} = \frac{1}{\mu_0} \overline{E} \times \overline{B} \quad \text{units: watts/meter}^2 \text{ (mks system)}$$

$\overline{S}$ = rate of energy

μ_0 = permeability constant

$\overline{E}$ = electric field

$\overline{B}$ = magnetic field

Problem Solving Example:

The solar constant, the power due to radiation from the sun falling on the earth's atmosphere, is 1.35 kW • m^{-2}. What are the magnitudes of $\overline{E}$ and $\overline{B}$ for the electromagnetic waves emitted from the sun at the position of the earth?

Starting with the electromagnetic waves at the earth, it is possible to determine $\overline{E}$ and $\overline{B}$ by two methods.

a)   The Poynting vector

$$\overline{S} = \frac{1}{\mu_0}\overline{E} \times \overline{B}$$

gives the energy flow across any section of the field per unit area per unit time.

Here, $\overline{E}$ and $\overline{B}$ are the instantaneous electric field and magnetic induction, respectively, at a point of space, and μ_0 is the permeability of free space. If we approximate the sun as a point source of light, then we realize that it radiates electromagnetic waves in all directions uniformly. However, the distance between earth and the sun is very large, and we may approximate the electromagnetic waves arriving at the surface of the earth as plane waves. For this type of wave, $\overline{E}$ and $\overline{B}$ are perpendicular. Thus,

$$\left|\overline{S}\right| = \left|\frac{1}{\mu_0}\overline{E} \times \overline{B}\right| = EH = 1.35 \times 10^3 \ \ W \bullet m^{-2}$$

where we have used the fact that $|\overline{H}| = \left|\dfrac{\overline{B}}{\mu_0}\right|$ in vacuum. (*H* is the magnetic field intensity.)

But in the electromagnetic field in vacuum,

$$\varepsilon_0 E^2 = \mu_0 H^2, \text{ or } E\sqrt{\frac{\varepsilon_0}{\mu_0}} = H.$$

Then

$$E \times \sqrt{\frac{\varepsilon_0}{\mu_0}} \ E = EH = 1.35 \times 10^3 \ \text{W} \cdot \text{m}^{-2}$$

or

$$E^2 = \sqrt{\frac{\mu_0}{\varepsilon_0}} (1.35 \times 10^3 \ \text{W} \cdot \text{m}^{-2})$$

$$= 377 \ \Omega \times 1.35 \times 10^3 \ \text{W} \cdot \text{m}^{-2}$$

$$E = \sqrt{5.09 \times 10^5} \ \text{V} \cdot \text{m}^{-1} = 7.1 \times 10^2 \text{V} \cdot \text{m}^{-1}$$

Similarly,

$$B = \mu_0 H = \mu_0 \frac{S}{E} = \left(4\pi \times 10^{-7} \ \frac{\text{Wb}}{\text{Am}}\right)\left(\frac{1.35 \times 10^3 \ \text{W/m}^{-2}}{7.1 \times 10^2 \ \text{V/m}}\right)$$

$$B = \frac{\left(4\pi \times 10^{-7} \text{Wb} \cdot \text{A}^{-1} \cdot \text{m}^{-1}\right)\left(1.35 \times 10^3 \ \text{W} \cdot \text{m}^{-2}\right)}{.71 \times 10^3 \ \text{V} \cdot \text{m}^{-1}}$$

$$B = 2.39 \times 10^{-6} \ \frac{\text{Wb} \cdot \text{W} \cdot \text{A}^{-1} \cdot \text{m}^{-2}}{\text{V}}$$

But $1 \ \text{W} = 1 \ \text{J} \cdot \text{s}^{-1}$ and $1 \ \text{V} = 1 \ \text{J} \cdot \text{C}^{-1}$ whence

$$B = 2.39 \times 10^{-6} \ \frac{\text{Wb} \cdot \text{J} \cdot \text{s}^{-1} \cdot \text{A}^{-1} \cdot \text{m}^{-2}}{\text{J} \cdot \text{C}^{-1}}$$

$$B = 2.39 \times 10^{-6} \ \text{Wb} \cdot \text{m}^{-2}$$

b) The electromagnetic energy density (or, energy per unit volume) in an electromagnetic field in vacuum is $\mu_0 H^2 = \varepsilon_0 E^2$. The energy falling on 1 m² of the earth's atmosphere in 1 s is the energy initially contained in a cylinder 1 m² in cross section and 3×10^8 m in length; for all this energy travels to the end of the cylinder in the space of 1 s. Hence, the energy density near the earth is

$$\mu_0 H^2 = \varepsilon_0 E^2 = \frac{(1.35 \times 10^3 \text{ W} \cdot \text{m}^{-2})}{3 \times 10^8 \text{ m} \cdot \text{s}^{-1}}$$

Here, ε_0 is the permittivity of free space.

$$E^2 = \frac{1.35 \times 10^3 \text{ W} \cdot \text{m}^{-2}}{(8.85 \times 10^{-12} \text{ C}^2 \cdot \text{N}^{-1} \cdot \text{m}^{-2})(3 \times 10^8 \text{ m} \cdot \text{s}^{-1})}$$

$$E^2 = \frac{1.35 \times 10^7}{26.55} \frac{\text{W}}{\text{C}^2 \cdot \text{N}^{-1} \cdot \text{m} \cdot \text{s}^{-1}}$$

But $1 \text{ W} = 1 \text{ J} \cdot \text{s}^{-1} = 1 \text{ N} \cdot \text{m} \cdot \text{s}^{-1}$

$$E^2 = 5.085 \times 10^5 \text{ N/C N}^{-1} = 5.085 \times 10^5 \text{ N}^2/\text{C}^2$$

or $E = 0.71 \times 10^3 \text{ N} \cdot \text{C}^{-1} = 0.71 \times 10^3 \text{ V} \cdot \text{m}^{-1}$

Also $\mu_0 H^2 = \dfrac{B^2}{\mu_0} = \dfrac{1.35 \times 10^3 \text{ W} \cdot \text{m}^{-2}}{3 \times 10^8 \text{ m} \cdot \text{s}^{-1}}$

or

$$B^2 = \frac{(4\pi \times 10^{-7} \text{ N} \cdot \text{A}^{-2})(1.35 \times 10^3 \text{ W} \cdot \text{m}^{-2})}{3 \times 10^8 \text{ m} \cdot \text{s}^{-1}}$$

$$B = 2.36 \times 10^{-6} \text{ Wb} \cdot \text{m}^{-2}$$

31.3 Polarization of Light

Intensity

$$I = I_m \cos^2 \theta$$

I = transmitted intensity

I_m = maximum transmitted intensity

θ = angle of polarization

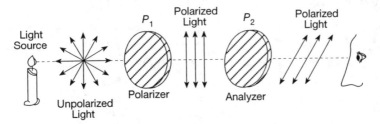

Figure 31.1 Light Polarization

31.4 The Doppler Effect for Light

$$v' = v \frac{1 - \dfrac{u}{c}}{\sqrt{1 - \left(\dfrac{u}{c}\right)^2}}$$

v' = observed frequency

v = actual frequency at source

u = speed of observer

c = speed of light

Note: The above equation shows the appropriate relations for the source and the observer separating; by substituting $(-u)$ for u, you will obtain the relations for the source and the observer approaching each other.

Problem Solving Example:

 As applied to visible light, the Doppler effect causes a light wave of a given frequency approaching the viewer to be perceived by objective instruments as being bluer. Why?

 Any approaching wave source is perceived as having a higher frequency. Therefore, light is bluer (not necessarily blue), i.e., closer to the blue end of the spectrum—higher frequency.

CHAPTER 32

Geometrical Optics

32.1 Reflection and Refraction

a) Law of Reflection

$$\theta_1' = \theta_1$$

b) Law of Refraction

$$\frac{\sin \theta_1}{\sin \theta_2} = n_{21}$$

θ_1' = angle of reflection

θ_1 = angle of incidence

θ_2 = angle of refraction

n_{21} = ratio of index of refraction of medium 2 with respect to the index or refraction of medium 1, i.e., $n_2/n_1 = n_{21}$

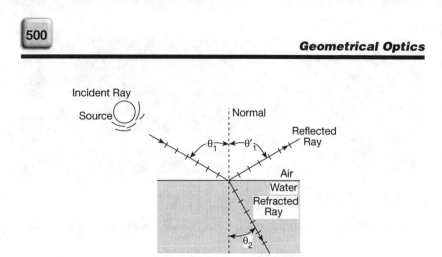

Incident Ray

Source

Normal

Reflected Ray

θ_1 θ'_1

Air
Water
Refracted Ray

θ_2

Figure 32.1 Reflection and Refraction

Problem Solving Example:

 An incident wavefront of light makes an angle of 60° with the surface of a pool of water. The speed of light in water is 2.3 × 10^8 m/s. What angle does the refracted wavefront make with the surface of water?

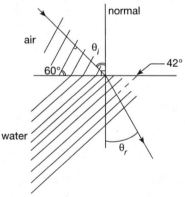

normal

air

θ_i

60°   42°

water

θ_r

The angle θ_i between the incident ray and the normal to the surface (as shown in the figure) equals the angle between the incident wavefront and the water surface, and

$$\theta_i = 60°.$$

Snell's Law, (see section 32.2) relating the angle of incidence, θ_i, to the angle of refraction, θ_r, of the light is

$$n_1 \sin \theta_i = n_2 \sin \theta_r$$

where n_1 and n_2 are the refractive indices of air and water, respectively. Hence,

$$\sin \theta_r = \frac{n_1}{n_2} \sin \theta_i$$

But

$$n_1 = \frac{\text{speed of light (vacuum)}}{\text{speed of light (air)}} \qquad n_2 = \frac{\text{speed of light (vacuum)}}{\text{speed of light (water)}}$$

Hence,

$$\sin \theta_r = \left(\frac{2.3 \times 10^8 \ \text{m/s}}{3 \times 10^8 \ \text{m/s}} \right) \sin 60° = 0.664$$

or
$$\theta_r = 42°$$

θ_r also equals the angle the refracted wavefront makes with the water surface.

32.2 Snell's Law

$$n_1 \sin \theta_1 = n_2 \sin \theta_2$$

n_1, n_2 = indices of refraction

a) $\quad v_n = \dfrac{c}{n}$

v_n = speed of light in the medium

c = speed of light in a vacuum

n = index of refraction

b) $\lambda_n = \dfrac{\lambda}{n}$

λ_n = wavelength in the medium

λ = wavelength in a vacuum

n = index of refraction

Problem Solving Example:

 A flat bottom swimming pool is 8 ft. deep. How deep does it appear to be when filled with water whose refractive index is 4/3?

 In order to see why we would expect to observe a different depth for the pool when it is filled with water, examine the figure.

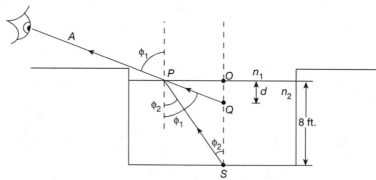

If no water is in the pool, light coming from a point S on the bottom of the pool will travel directly to the observer's eye. If the pool is filled with water, light emanating from point S will be refracted at P, as shown. Upon reaching the observer's eye, the light appears to be coming from Q and he or she perceives the depth of the pool to be the distance OQ, rather than the actual depth OS. Our problem is to find the distance d.

Note that, from the figure,

$$\tan \varphi_1 = \frac{OP}{d}$$

$$\tan \varphi_2 = \frac{OP}{8 \text{ ft.}}$$

Hence,

$$\frac{\tan \varphi_1}{\tan \varphi_2} = \frac{OP}{d} \cdot \frac{8 \text{ ft.}}{OP} = \frac{8 \text{ ft.}}{d}$$

and

$$d = \frac{(8 \text{ ft}) \tan \varphi_2}{\tan \varphi_1} \tag{1}$$

From Snell's Law,

$$n_1 \sin \varphi_1 = n_2 \sin \varphi_2$$

where n_1 and n_2 are the indices of refraction of air and water, respectively. Therefore,

$$\left(\frac{n_1}{n_2}\right) \sin \varphi_1 = \sin \varphi_2 \tag{2}$$

To calculate the tangents in (1), we must also know $\cos \varphi_1$ and $\cos \varphi_2$. These we may find by observing that

$$\cos \varphi = \sqrt{1 - \sin^2 \varphi} \tag{3}$$

Using (2) in (3)

$$\cos \varphi_2 = \sqrt{1 - \sin^2 \varphi_2}$$

$$\cos \varphi_2 = \sqrt{1 - \left(\frac{n_1}{n_2}\right)^2 \sin^2 \varphi_1} \tag{4}$$

$$\cos \varphi_1 = \sqrt{1 - \sin^2 \varphi_1}$$

Hence,

$$\tan \varphi_1 = \frac{\sin \varphi_1}{\cos \varphi_1} = \frac{\sin \varphi_1}{\sqrt{1 - \sin^2 \varphi_1}} \tag{5}$$

and using (2) with (4),

$$\tan \varphi_2 = \frac{\sin \varphi_2}{\cos \varphi_2} = \frac{(n_1 / n_2) \sin \varphi_1}{\sqrt{1 - (n_1 / n_2)^2 \sin^2 \varphi_1}} \tag{6}$$

Substituting (5) and (6) in (1),

$$d = (8 \text{ ft}) \ \frac{(n_1 / n_2) \ \sin \varphi_1}{\sqrt{1 - (n_1 / n_2)^2 \sin^2 \varphi_1}} \cdot \frac{\sqrt{1 - \sin^2 \varphi_1}}{\sin \varphi_1}$$

$$d = (8 \text{ ft}) \left(\frac{n_1}{n_2} \right) \sqrt{\frac{1 - \sin^2 \varphi_1}{1 - (n_1 / n_2)^2 \sin^2 \varphi_1}} \tag{7}$$

Now, since we don't know the angle φ_1, we make an approximation. Suppose φ_1 is very small. (This means that the observer is looking almost directly down into the pool.) Then $\sin \varphi_1 \approx 0$ and the square root in (7) becomes 1. Therefore,

$$d = (8 \text{ ft}) \left(\frac{n_1}{n_2} \right) = (8 \text{ ft}) \left(\frac{1}{4 / 3} \right) = 6 \text{ ft.}$$

The pool appears to be 6 ft. deep.

32.3 Total Internal Reflection

Critical Angle, θ_c

$$\sin \theta_c = \frac{n_2}{n_1}$$

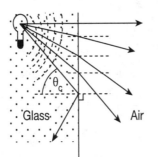

θ_c is the critical angle
(total internal reflection)

Glass   Air

Figure 32.2 Internal Reflection

Problem Solving Example:

Q Describe the phenomena of the critical angle in optics. What is the critical angle for a glass-air interface if the index of refraction of glass with respect to air is 1.33?

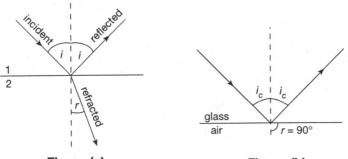

Figure (a)   **Figure (b)**

A Consider two media, 1 and 2, such that the index of refraction of 1 with respect to 2 is greater than unity, as shown in Figure (a). That is, medium 1 is "denser," and the angle of refraction will be greater than the angle of incidence.

In general, part of the incident ray is reflected and part refracted. As the angle of incidence is increased, the angle of refraction will increase until $r = 90°$. At this critical angle, i_c, (see Figure (b)), we have

$$\frac{\sin i_c}{\sin 90°} = \sin i_c = n_{21}$$

At the critical angle, and for values of i greater than i_c, refraction cannot occur and all the energy of the incident beam appears in the reflected beam. This phenomenon is called total internal reflection.

The index of refraction of air with respect to glass is

$$\frac{1}{1.33} = 0.75.$$

The critical angle for a glass-air interface is therefore

$$\sin i_c = 0.75$$

or

$$i_c = 48.6°$$

Thus, for angles of incidence $\geq 48.6°$, total internal reflection will occur for a glass-air combination similar to the one shown in the figure on the previous page.

32.4 Brewster's Law

$$\tan \theta_p = \frac{n_2}{n_1}$$

θ_p = polarizing angle

32.5 Incident Angle Plane Mirrors

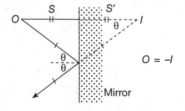

Figure 32.3 Geometry of a Plane Mirror

32.6 Spherical Mirrors

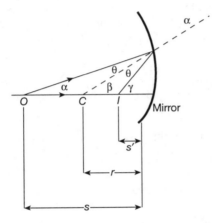

Figure 32.4 Geometry of a Spherical Mirror

Spherical Mirror

$$\frac{1}{s} + \frac{1}{s'} = \frac{2}{r}$$

s = object distance

s' = image distance

r = radius of curvature

$$\frac{1}{2}r = f$$

r = radius of curvature

f = focal length

$$\frac{1}{s} + \frac{1}{s'} = \frac{1}{f}$$

s = object distance

s' = image distance

f = focal length

$$m = \frac{h'}{h}$$

m = magnification factor

h' = image height

h = object height

O = object at distance, S

I = image at distance, S'

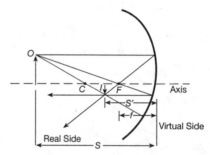

(a) Concave mirror, inverted image

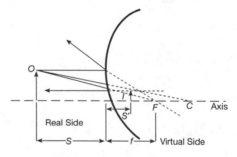

(b) Convex mirror, erect image

Figure 32.5 Objects and Images on Spherical Mirrors

F = focal point

C = center of curvature

f = focal length

Problem Solving Example:

 An object is 12 feet from a concave mirror whose focal length is 4 feet. Where will the image be found?

We first construct a ray diagram.

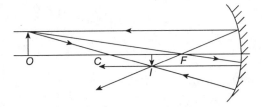

The image is a real image (inverted) between the center of curvature and the focal point. It is smaller than the object.

We now solve mathematically:

s_0 is 12 feet and f is 4 feet. Substituting in

$$\frac{1}{s_0} + \frac{1}{s_I} = \frac{1}{f}$$

$$\frac{1}{12 \text{ ft.}} + \frac{1}{s_I} = \frac{1}{4 \text{ ft.}}$$

whence $\qquad\qquad s_r = 6 \text{ ft.}$

which means that since s_I is negative the image is 6 feet from the mirror on the same side on the mirror as the object and is real.

32.7 Spherical Refracting Surface

$$\frac{n_1}{s} + \frac{n_2}{s'} = \frac{n_2 - n_1}{r}$$

For $\alpha \approx 0°$

$\quad$ n = index of refraction

$\quad$ s = object distance

$\quad$ s' = image distance

$\quad$ r = radius of curvature

32.8 Thin Lenses

a) $\quad$ Refraction by a Thin Lens

$$\frac{1}{s} + \frac{1}{s'} = (n-1)\left(\frac{1}{r'} - \frac{1}{r''}\right)$$

$\quad$ s = object distance

$\quad$ s' = image distance

$\quad$ n = index of refraction

$\quad$ r' = radius of curvature of first surface struck by light

$\quad$ r'' = radius of curvature of second surface struck by light

b) $\quad$ Focal Points of a Thin Lens

$$\frac{1}{f} = (n-1)\left(\frac{1}{r'} - \frac{1}{r''}\right)$$

$\quad$ where f = focal length

c) $\quad$ Thin Lens Equation

$$\frac{1}{s} + \frac{1}{s'} = \frac{1}{f}$$

d) Magnification of a Thin Lens (Linear)

$$m = \frac{-s'}{s}$$

32.9 Thin Lens Diagrams

a) Convex Lens with Real Image

F_1, F_2 = focal points

f = focal length

b) Convex Lens with Virtual Image

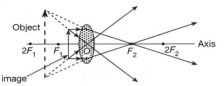

c) Concave Lens with Virtual Image

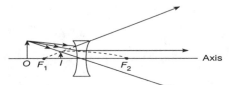

Figure 32.6 Graphical Locations of Images for Three Different Object Distances and Three Thin Lenses

Problem Solving Examples:

 If an object is in a position 8 cm in front of a lens of focal length 16 cm, where and how large is the image? (See figure.)

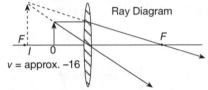

Ray Diagram

v = approx. −16

 We refer here to the relationship

$$\frac{1}{s_0} + \frac{1}{s_I} = \frac{1}{f}$$

Here, the image distance, i, is the unknown. The object distance, 0, is 8 cm, while the focal length, f, is 16 cm. Thus,

$$\frac{1}{s_0} + \frac{1}{s_i} = \frac{1}{f}$$

$$\frac{1}{8 \text{ cm}} + \frac{1}{s_i} = \frac{1}{16 \text{ cm}}$$

$$\frac{1}{s_i} = \frac{1}{16 \text{ cm}} - \frac{1}{8 \text{ cm}}$$

$$\frac{1}{s_i} = -\frac{1}{16 \text{ cm}}$$

$$s_i = -16 \text{ cm}$$

This means that the image is 16 cm in front of the lens.

To calculate the image size, we first calculate the magnification. Thus,

$$m = \frac{-i}{0} = \frac{+16}{8} = 2$$

This means that the image is erect, therefore virtual, and twice as large as the object.

 An object is 4 inches from a concave lens whose focal length is –12 inches. Where will the image be?

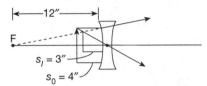

 It may be useful to construct a ray-diagram first (see diagram). We draw two rays—one parallel to the axis and one through the center of the lens.

From the diagram it can be seen that the image is virtual and that it is smaller than the object.

We now attempt a mathematical solution:

s_0 is 4 inches and F is –12 inches. Substituting in

$$\frac{1}{s_0} + \frac{1}{s_I} = \frac{1}{F}$$

$$\frac{1}{4 \text{ in.}} + \frac{1}{s_I} = \frac{1}{-12 \text{ in.}}$$

Solving $s_I = -3$ in. Which means that since s_I is negative, the image is 3 inches from the lens on the same side on the lens as the object and is virtual.

Quiz: Electromagnetic Waves— Geometrical Optics

QUESTIONS 1 and 2 refer to the following diagram.

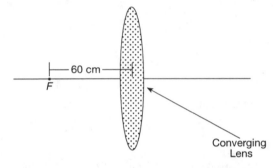

60 cm

F

Converging
Lens

1. An object is placed 30 centimeters to the left of the lens shown. Where will the image of the object be located?

    (A) 60 cm to the left of the lens

    (B) 20 cm to the left of the lens

    (C) 20 cm to the right of the lens

    (D) 30 cm to the right of the lens

    (E) 60 cm to the right of the lens

2. The image formed will be

    (A) virtual and reduced.   (D) real and enlarged.

    (B) virtual and enlarged.   (E) real and inverted.

    (C) real and reduced.

3. All of the following phenomena can be explained if light is a wave EXCEPT

(A) reflection.

(D) diffraction.

(B) refraction.

(E) interference.

(C) photoelectric effect.

4. Images produced by a plane mirror are

(A) virtual, erect, and the same size.

(B) virtual, inverted, and larger.

(C) virtual, inverted, and smaller.

(D) real, inverted, and larger.

(E) real, erect, and smaller.

5. The reflected image of an object in front of a plane mirror always appears

(A) smaller than the object.

(B) the same size as the object and inverted.

(C) as far behind the mirror as the object is in front of the mirror.

(D) distorted.

(E) diffused.

6. Refraction occurs when

(A) a wave enters a new medium at some angle.

(B) a wave bends around a barrier.

(C)   a wave bounces back on itself.

(D)   a wave runs into another wave.

(E)   a wave changes speed at a boundary of a new medium.

7.   A convex lens that converges parallel light rays creates an image that is real when the object is

(A)   at the focal length of the lens.

(B)   beyond the focal length of the lens.

(C)   on the same side of the lens as the image.

(D)   between the focal length of the lens and the lens.

(E)   virtual.

8.   The index of refraction for an unknown substance is 1.2. The critical angle for total internal reflection to occur in the substance if going from the substance to air is found by

(A) $\sin(1.2) = \theta$.   (D) $\sin\theta = 0.8$.

(B) $\sin(0.8) = \theta$.   (E) $\dfrac{1}{\sin}(1.2) = \theta$.

(C) $\sin\theta = 1.2$.

9.   A wave leaves one medium in which it travels 6 m/s and has a wavelength of 2 m and enters a second medium in which the wavelength is 3 m. The wave speed in the second wave is

(A)   1.5 m/s.   (D)   9 m/s.

(B)   5 m/s.   (E)   18 m/s.

(C)   6 m/s.

10. If a segment, the distance between two nodes in a standing wave, is 20 cm long, the wavelength of the original traveling wave used to create the standing wave is

(A) 10 cm.   (D) 40 cm.

(B) 20 cm.   (E) 50 cm.

(C) 30 cm.

ANSWER KEY

1.	(A)	6.	(E)
2.	(B)	7.	(B)
3.	(C)	8.	(C)
4.	(A)	9.	(D)
5.	(C)	10.	(D)

CHAPTER 33

Interference

33.1 Young's Experiment

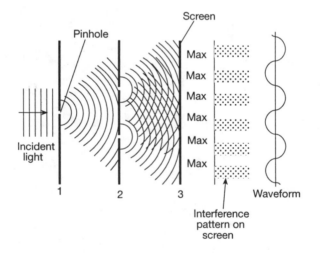

Figure 33.1 Diffraction Wave Patterns for Young's Double-slit Experiment

33.2 Interference Maxima and Minima

a) Maxima

$$d \sin \theta = m\lambda$$

where $m = 0,1,2,\ldots$ (maxima)

b) Minima

$$d \sin \theta = \left(m + \frac{1}{2}\right)\lambda$$

where $m = 1, 2,\ldots$ (minima)

d = distance between slits

θ = angle between the midpoint of the two slits and a point on the screen

λ = wavelength

Problem Solving Example:

 In a double-slit experiment, $D = 0.1$ mm and $L = 1$ m. If yellow light is used, what will be the spacing between adjacent bright lines?

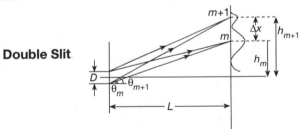

Double Slit

 The wavelength of yellow light is approximately 6×10^{-5} cm. Let the separation between the $(m + 1)$'th and m'th maxima be Δx, as shown in the figure. Then,

$$\sin \theta_m = \frac{m}{D}\lambda$$

$$\sin \theta_{m+1} = \frac{m+1}{D}\lambda.$$

The angles θ_m and θ_{m+1} are related to the positions h_m and h_{m+1} of these maxima on the screen

$$\sin \theta_m \cong \frac{h_{m+1}}{L}$$

$$\sin \theta_{m+1} \cong \frac{h_m}{L}$$

hence,

$$\sin \theta_{m+1} - \sin \theta_m = \frac{\lambda}{D}$$

$$\frac{h_{m+1} - h_m}{L} \cong \frac{\lambda}{D}$$

or

$$\frac{\Delta x}{L} \approx \frac{\lambda}{D}$$

$$\Delta x \cong \frac{L}{D}$$

$$= \frac{(6 \times 10^{-5} \text{cm}) \times (100 \text{ cm})}{10^{-2} \text{cm}}$$

$$= 0.6 \text{ cm}$$

Thus, the spacing between lines is about 6 mm or $1/4$ of an inch.

33.3 Adding Phasors

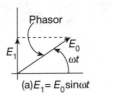
(a)$E_1 = E_0 \sin \omega t$

+

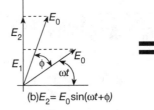
(b)$E_2 = E_0 \sin(\omega t + \phi)$

=

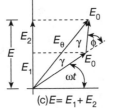
(c)$E = E_1 + E_2$

(d) $E = E_\theta \sin(\omega t + \gamma)$

Note: $E_{\text{TOTAL}} = (E_1 + E_2 + E_3 + \ldots + E_n)$

Figure 33.2 Phasor (Rotating Vector) Addition

33.4 Thin-film Interference

Interference Maxima and Minima

$$2(n)(d) = (m + \frac{1}{2})\lambda \quad \text{Constructive Interference}$$

where $m = 0, 1, 2, \ldots$(maxima)

$$2(n)(d) = m\lambda \quad \text{Destructive Interference}$$

where $m = 0, 1, 2, \ldots$(minima)

n = index of refraction of film material with respect to the medium

d = thickness of medium

λ = wavelength

Problem Solving Example:

Q When a flat plate of glass and a lens are placed in contact, a distinctive interference pattern, known as Newton's Rings, is observed. (See Figure (a).) Derive a formula giving the location of the fringes of the interference pattern relative to the center of the lens. (See Figure (b).)

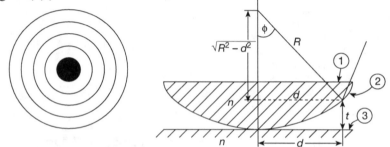

Figure (a): Interference Pattern (Top View)

Figure (b)

A Destructive interference will result when the waves reflected from the apparatus shown in Figure (b) are 180° out of phase. Let us trace the path of an incident ray. The ray will be partially reflected and partially transmitted at surface (1). When a ray of light is transmitted from a region of low refractive index to a medium of higher refractive index, it undergoes a phase change of 180°. Hence, the ray reflected at surface (1) is 180° out of phase with the incident ray. The transmitted ray next encounters surface (2). At this surface there is no phase change, since the light leaves an area of high refractive index and enters a region of low refractive index. In addition, part of this light is reflected at surface (2). The light transmitted at surface (2) next encounters surface (3) and is reflected with a 180° or $\lambda/2$ phase change. (λ is the wavelength of the light.) Hence, the ray reflected at (2) and the ray reflected at (3) are 180° out of phase.

Now, the ray reflected at (3) travels a distance $2t$ greater than the ray reflected at (2). We will see destructive interference whenever $2t$ is an integral number of wavelengths, since the additional $\lambda/2$ required for destruction is provided by the phase change due to reflection. Hence,

$$2t = n\lambda \qquad (1)$$

$n = 0, 1, 2, \ldots$ destructive interference

$$2t = \left(n + \frac{1}{2}\right)\lambda$$

$n = 0, 1, 2, \ldots$ .constructive interference

We must now find the location of the interference fringes in terms of the geometry of Figure (b) on the previous page.

From Figure (b),

$$t = R - \sqrt{R^2 - d^2} \qquad (2)$$

$$t = R - R\sqrt{1 - \frac{d^2}{R^2}}$$

$$t = R\left\{1 - \sqrt{1 - \frac{d^2}{R^2}}\right\} \qquad (3)$$

But $d \ll R$ and $d/R \ll 1$. (This means that the radius of curvature of the lens is large.) We may therefore approximate the square root in (3) by the binomial theorem. Therefore,

$$\left(1 - \frac{d^2}{R^2}\right)^{1/2} \approx 1 - \frac{1}{2}\frac{d^2}{R^2} \qquad (4)$$

Substituting (4) in (3),

$$t = R\left\{1 - \left(1 - \frac{d^2}{2R^2}\right)\right\}$$

$$t = R\left\{1 - 1 + \frac{d^2}{2R^2}\right\}$$

$$t = \frac{Rd^2}{2R^2} = \frac{d^2}{2R}$$

Using (1),

$$d = \sqrt{n\lambda R}$$

$n = 0,1,2,...$ destructive interference

$$d = \sqrt{\left(n + \frac{1}{2}\right)\lambda R}$$

$n = 0,1,2,...$ constructive interference

The first equation locates the dark rings relative to the center of the lens, and the second equation locates the bright rings.

CHAPTER 34

Diffraction

34.1 Single-slit Diffraction

a)

$$a \, \sin \theta = m\lambda$$

$m = 1, 2, 3, \ldots$ (minima)

a = width of slit

θ = angle of diffraction

λ = wavelength

b) Diffracted Intensity

$$I_\theta = I_m \left(\frac{\sin \alpha}{\alpha} \right)^2$$

where

$$\alpha = \frac{\pi a}{\lambda} \sin \theta$$

I_θ = diffracted intensity

I_m = maximum intensity

a = width of slit

λ = wavelength

θ = angle of diffraction

Problem Solving Example:

Q A slit of width a is placed in front of a lens of focal length 50 cm and is illuminated normally with light of wavelength 5.89×10^{-5} cm. The first minima on either side of the central maximum of the diffraction pattern observed in the focal plane of the lens are separated by 0.20 cm. What is the value of a?

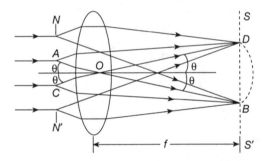

A This is an example of Fraunhofer diffraction (see figure). Parallel rays of light are incident on slit NN' from the left. The rays are diffracted and encounter the lens of focal length f. The rays are then focused on a screen (SS') lying in the lens' focal plane, and a diffraction pattern is observed. The minima of this diffraction pattern are described by the formula

$$\sin\theta = \frac{m\lambda}{a}$$

where θ locates the minima fringes, a is the slit width, m is the minima number, and λ is the wavelength of the light used. In this problem, the first minima on either side of the central maximum is found at angle θ, such that

$$\theta = \sin^{-1}\frac{\lambda}{a}.$$

The angular separation of the two minima is then (see figure on previous page):

$$\delta = 2\theta = 2 \ \sin^{-1}\frac{\lambda}{a} \tag{1}$$

But, we may write

$$2\theta \approx \frac{\overline{DB}}{f} \tag{2}$$

where DB is the linear separation of the two minima. Here, we have approximated the arc DB (shown dotted in the figure) by the linear distance $\overline{DB}$. Using (2) in (1)

$$\frac{\overline{DB}}{f} \approx 2\sin^{-1}\frac{\lambda}{a}$$

or

$$\sin\left(\frac{\overline{DB}}{2f}\right) \approx \frac{\lambda}{a}$$

and

$$a \approx \frac{\lambda}{\sin\left(\dfrac{\overline{DB}}{2f}\right)}$$

Hence,

$$a \approx \frac{5.89 \times 10^{-5} \ \text{cm}}{\sin\left(\dfrac{0.20 \ \text{cm}}{100 \ \text{cm}}\right)} = \frac{5.89 \times 10^{-5} \ \text{cm}}{\sin (0.0020)}$$

Since 0.0020 rad is a very small angle, we may write

$$\sin (0.0020) \approx 0.0020 = 2 \times 10^{-3}$$

whence $\qquad a \approx \dfrac{5.89 \times 10^{-5} \ \text{cm}}{2 \times 10^{-3}} = 2.945 \times 10^{-2} \ \text{cm}$

34.2 Circular Diffraction

$$\sin\theta = 1.22\frac{\lambda}{d}$$

θ = angle of diffraction

λ = wavelength

d = diameter of circular aperture or lens

34.3 Rayleigh's Criterion

$$\theta_r = 1.22\frac{\lambda}{d}$$

θ_r = angular separation

λ = wavelength

d = diameter of circular aperture or lens

34.4 Double-slit Diffraction

$$I_\theta = I_m(\cos\beta)^2\left(\frac{\sin\alpha}{\alpha}\right)^2$$

where $\beta = \dfrac{\pi d}{\lambda}\sin\theta$ and $\alpha = \dfrac{\pi a}{\lambda}\sin\theta$

I_θ = diffracted intensity

I_m = maximum intensity

d = distance between centers of slits

a = width of slit

λ = wavelength

θ = angle of diffraction

Problem Solving Example:

Q The deviation of the second-order diffracted image formed by an optical grating having 5,000 lines/cm is 32°. Calculate the wavelength of the light used.

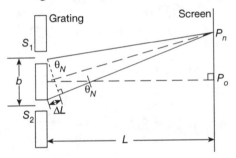

A An optical grating is a transparent piece of glass or plastic on which there are closely spaced parallel scratches. These scratches behave like opaque barriers and the spaces between the scratches like narrow slits. Each adjacent pair of slits produces a double-slit diffraction pattern so that this problem can be analyzed using double-slit interference.

Point P_n in the diagram receives light from both slits. The path lengths from slits S_1 and S_2 to P_n differ by an amount very close to ΔL. If the two waves are in phase as they leave slits S_1 and S_2, then they will also be in phase at P_n if ΔL is an integral multiple of the wavelength of the light. That is, if $\Delta L = n\lambda$ then reinforcement occurs at P_n and a bright band occurs there. If b is small compared to L, then the angle θ

in the small right triangle is nearly equal to θ in the larger right triangle. Therefore, we say

$$\sin \theta_N = \frac{N\lambda}{b}$$

or

$$b \sin \theta_N = N\lambda$$

where θ_N is the deviation of the Nth order diffracted image and b is the distance between slits.

$$b = \frac{1}{5,000 \, \text{cm}^{-1}} = 0.00020 \ \text{cm}$$

$$\lambda = \frac{b \sin \theta_N}{N} = \frac{0.00020 \ \text{cm} \times 0.53}{2}$$

$$= 0.000053 \ \text{cm} = 5,300 \, \text{Å}$$

34.5 Multiple-slit Diffraction

a) Maxima

$$d \sin \theta = m\lambda$$

$m = 0, 1, 2,...$

d = distance between centers of slits

θ = angle of diffraction

λ = wavelength

b) Angular Width

$$\Delta\theta_m = \frac{\lambda}{Nd \cos \theta_m}$$

$\Delta\theta_m$ = angular width of the maxima

λ = wavelength

N = number of slits

d = distance between the centers of slits

θ_m = angle of diffraction from the principal maximum

Problem Solving Example:

The spectrum of a particular light source consists of lines and bands stretching from a wavelength of 5.0×10^{-5} cm to 7.5×10^{-5} cm. When a diffraction grating is illuminated normally with this light it is found that two adjacent spectra formed just overlap, the junction of the two spectra occurring at an angle of 45°. How many lines per centimeter are ruled on the grating?

The grating formula is

$$d \sin\theta = n\lambda$$

where λ is the wavelength of light incident upon the grating, d is the grating spacing, n is the order number, and θ locates the maxima of the diffraction pattern. At the angle of 45°, we have $d \sin 45° = m \times 7.5 \times 10^{-5}$ cm, and also $d \sin 45° = (m + 1) \times 5.0 \times 10^{-5}$ cm. (We can see why the smaller wavelength has the larger order number by examining the grating formula, $d \sin\theta = n\lambda$. Since θ and d are the same for both λ's, we obtain $n\lambda$ = const. Hence, at a particular θ, the larger the wavelength, the smaller must be n, and vice-versa.)

$$\therefore \quad \frac{m+1}{m} = \frac{7.5}{5.0} = \frac{3}{2}. \quad \therefore \quad m = 2.$$

The second-order spectrum thus just overlaps with the third. Also, using the first formula above,

$$d = \frac{2 \times 7.5 \times 10^{-5} \text{ cm}}{\sin 45°} = 2.12 \times 10^{-4} \text{ cm}.$$

This is the separation of the rulings. Hence, the number of rulings per centimeter, n, is

$$n = \frac{1}{d} = \frac{10^4}{2.12 \ \text{cm}} = 4,715 \ \text{per cm.}$$

34.6 Diffraction Gratings

a) Dispersion

$$D = \frac{m}{d \cos \theta}$$

D = dispersion (angular separation)

m = order of maxima

θ = angle of diffraction

b) Resolving Power

$$R = \frac{\lambda}{\Delta\lambda} = Nm$$

R = resolving power

λ = mean wavelength

$\Delta\lambda$ = wavelength difference

N = number of rulings in the grating

m = order of maxima

34.7 X-ray Diffraction

Bragg's Law

$$2d \sin \theta = m\lambda$$

$m = 1, 2, 3,\ldots$

d = inter-planar spacing

θ = angle of diffraction

λ = wavelength

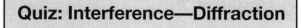

Quiz: Interference—Diffraction

QUESTIONS 1–3 refer to the following wave phenomena.

   I. Diffraction

   II. Interference

   III. Reflection

1. Occurs when two or more waves pass through the same region.

   (A) I   (D) I and III

   (B) II   (E) II and III

   (C) III

2. Standing waves are produced when periodic waves experience this phenomenon.

   (A) I   (D) I and III

   (B) II   (E) II and III

   (C) III

3. The spreading of a wave into a region behind an object is said to experience this phenomenon.

   (A) I   (D) I and III

   (B) II   (E) II and III

   (C) III

QUESTION 4 refers to the following diagram.

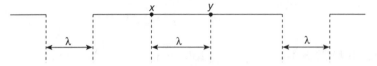

4. Two wave pulses, each of wavelength λ, are traveling toward each other along a rope as shown above. When both pulses are in the region between points *x* and *y* and which are λ distance apart, the shape of the rope could be which of the following?

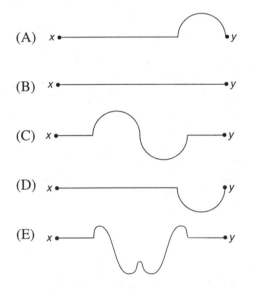

5. As the wavelength of a wave increases, the amount of diffraction the wave experiences in going around a barrier

(A) increases.   (D) cancels.

(B) decreases.   (E) interferes.

(C) remains the same.

QUESTIONS 6–8 refer to the following diagrams.

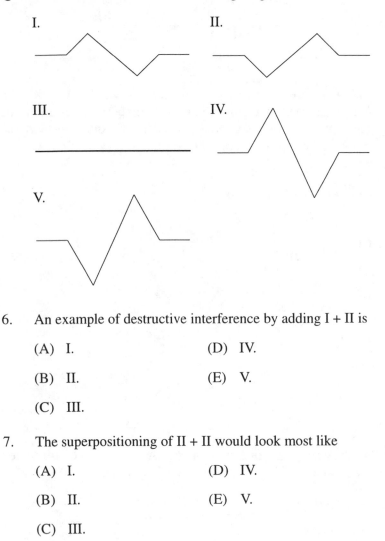

I.

II.

III.

IV.

V.

6.   An example of destructive interference by adding I + II is

(A) I.   (D) IV.

(B) II.   (E) V.

(C) III.

7.   The superpositioning of II + II would look most like

(A) I.   (D) IV.

(B) II.   (E) V.

(C) III.

8. The superpositioning of V + I would look most like

    (A) I.   (D) IV.

    (B) II.   (E) V.

    (C) III.

9. Consider a Young double-slit experiment where the two slits are spaced $d = 0.1$ mm apart. If when the screen is at a distance $l = 1$ m, the first bright maximum is displaced $y = 2$ cm from the central maximum. Find the wavelength of the light.

    (A) 4,000 Å   (D) 5,000 Å

    (B) 8,000 Å   (E) 2,000 Å

    (C) 10,000 Å

10. When monochromatic light passes through a double slit, a bright fringe is formed due to a path difference of

    (A) $\frac{1}{4}\lambda$.   (D) $\frac{7}{8}\lambda$.

    (B) $\frac{1}{2}\lambda$.   (E) $1\,\lambda$.

    (C) $\frac{3}{4}\lambda$.

ANSWER KEY

1.	(B)		6.	(C)
2.	(E)		7.	(E)
3.	(A)		8.	(B)
4.	(B)		9.	(E)
5.	(A)		10.	(E)

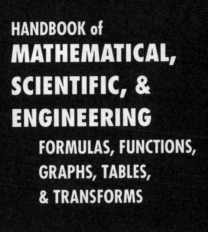

REA's **Problem Solvers**

The "PROBLEM SOLVERS" are comprehensive supplemental text-books designed to save time in finding solutions to problems. Each "PROBLEM SOLVER" is the first of its kind ever produced in its field. It is the product of a massive effort to illustrate almost any imaginable problem in exceptional depth, detail, and clarity. Each problem is worked out in detail with a step-by-step solution, and the problems are arranged in order of complexity from elementary to advanced. Each book is fully indexed for locating problems rapidly.

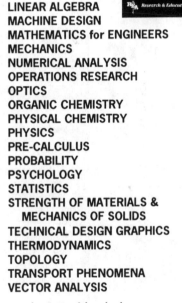

ACCOUNTING
ADVANCED CALCULUS
ALGEBRA & TRIGONOMETRY
AUTOMATIC CONTROL
 SYSTEMS/ROBOTICS
BIOLOGY
BUSINESS, ACCOUNTING, & FINANCE
CALCULUS
CHEMISTRY
COMPLEX VARIABLES
DIFFERENTIAL EQUATIONS
ECONOMICS
ELECTRICAL MACHINES
ELECTRIC CIRCUITS
ELECTROMAGNETICS
ELECTRONIC COMMUNICATIONS
ELECTRONICS
FINITE & DISCRETE MATH
FLUID MECHANICS/DYNAMICS
GENETICS
GEOMETRY
HEAT TRANSFER

LINEAR ALGEBRA
MACHINE DESIGN
MATHEMATICS for ENGINEERS
MECHANICS
NUMERICAL ANALYSIS
OPERATIONS RESEARCH
OPTICS
ORGANIC CHEMISTRY
PHYSICAL CHEMISTRY
PHYSICS
PRE-CALCULUS
PROBABILITY
PSYCHOLOGY
STATISTICS
STRENGTH OF MATERIALS &
 MECHANICS OF SOLIDS
TECHNICAL DESIGN GRAPHICS
THERMODYNAMICS
TOPOLOGY
TRANSPORT PHENOMENA
VECTOR ANALYSIS

If you would like more information about any of these books,
complete the coupon below and return it to us or visit your local bookstore.

RESEARCH & EDUCATION ASSOCIATION
61 Ethel Road W. • Piscataway, New Jersey 08854
Phone: (732) 819-8880   **website: www.rea.com**

Please send me more information about your Problem Solver books

Name _____

Address _____

City _____ State _____ Zip _____

REA's Test Preps
The Best in Test Preparation

- REA "Test Preps" are **far more** comprehensive than any other test preparation series
- Each book contains up to **eight** full-length practice tests based on the most recent exam
- **Every** type of question likely to be given on the exams is included
- Answers are accompanied by **full** and **detailed** explanations

REA publishes over 60 Test Preparation volumes in several series. They include:

Advanced Placement Exams (APs)
Biology
Calculus AB & Calculus BC
Chemistry
Computer Science
English Language & Composition
English Literature & Composition
European History
Government & Politics
Physics
Psychology
Spanish Language
Statistics
United States History

College-Level Examination Program (CLEP)
Analyzing and Interpreting Literature
College Algebra
Freshman College Composition
General Examinations
General Examinations Review
History of the United States I
Human Growth and Development
Introductory Sociology
Principles of Marketing
Spanish

SAT II: Subject Tests
Biology E/M
Chemistry
English Language Proficiency Test
French
German
Literature

SAT II: Subject Tests (cont'd)
Mathematics Level IC, IIC
Physics
Spanish
United States History
Writing

Graduate Record Exams (GREs)
Biology
Chemistry
Computer Science
General
Literature in English
Mathematics
Physics
Psychology

ACT - ACT Assessment

ASVAB - Armed Services Vocational
Aptitude Battery

CBEST - California Basic Educational
Skills Test

CDL - Commercial Driver License Exam

CLAST - College-Level Academic Skills Test

ELM - Entry Level Mathematics

ExCET - Exam for the Certification of
Educators in Texas

FE (EIT) - Fundamentals of Engineering
Exam

FE Review - Fundamentals of
Engineering Review

GED - High School Equivalency
Diploma Exam (U.S. & Canadian
editions)

GMAT - Graduate Management
Admission Test

LSAT - Law School Admission Test

MAT - Miller Analogies Test

MCAT - Medical College Admission
Test

MTEL - Massachusetts Tests for
Educator Licensure

MSAT - Multiple Subjects Assessment for Teachers

NJ HSPA - New Jersey High School
Proficiency Assessment

PLT - Principles of Learning &
Teaching Tests

PPST - Pre-Professional Skills Tests

PSAT - Preliminary Scholastic
Assessment Test

SAT I - Reasoning Test

SAT I - Quick Study & Review

TASP - Texas Academic Skills
Program

TOEFL - Test of English as a
Foreign Language

TOEIC - Test of English for
International Communication

RESEARCH & EDUCATION ASSOCIATION
61 Ethel Road W. • Piscataway, New Jersey 08854
Phone: (732) 819-8880   **website: www.rea.com**

Please send me more information about your Test Prep books

Name _____

Address _____

City _____ State _____ Zip _____